MEDIA AND COMMUNICATIONS - TECHNOLOGIES, POLICIES AND CHALLENGES

PUBLIC SAFETY BROADBAND AND COMMUNICATION NETWORKS

PLANS AND CONSIDERATIONS

Media and Communications - Technologies, Policies and Challenges

Additional books in this series can be found on Nova's website under the Series tab.

Additional E-books in this series can be found on Nova's website under the E-book tab.

Computer Networks

Additional books in this series can be found on Nova's website under the Series tab.

Additional E-books in this series can be found on Nova's website under the E-book tab.

MEDIA AND COMMUNICATIONS - TECHNOLOGIES, POLICIES AND CHALLENGES

PUBLIC SAFETY BROADBAND AND COMMUNICATION NETWORKS

PLANS AND CONSIDERATIONS

THOMAS GIORDANO
EDITOR

New York

For permission to use material from this book please contact us:
Telephone 631-231-7269; Fax 631-231-8175
Web Site: http://www.novapublishers.com

Additional color graphics may be available in the e-book version of this book.

LIBRARY OF CONGRESS CATALOGING-IN-PUBLICATION DATA

ISBN: 978-1-62417-524-4

Published by Nova Science Publishers, Inc. † New York

CONTENTS

PREFACE

Since September 11, 2001, when communication failures contributed to the tragedies of the day, Congress has passed several laws intended to create a nationwide emergency communications capability. The United States has continued to strive for a solution that assures seamless communications among first responders and emergency personnel at the scene of a major disaster. To address this problem, Congress included provisions in the Middle Class Tax Relief and Job Creation Act of 2012, for planning, building, and managing a new, nationwide, broadband network for public safety communications, and assigned additional spectrum to accommodate the new network. This book explores public safety broadband and communication networks with a focus on new policies for spectrum management and wireless innovation that would facilitate the transition to IP-enabled networks. Acceleration of innovation in next-generation wireless technologies would likely benefit not only public safety communications but also all consumers of wireless service and the American economy.

Chapter 1 – Since September 11, 2001, when communications failures contributed to the tragedies of the day, Congress has passed several laws intended to create a nationwide emergency communications capability. Yet the United States has continued to strive for a solution that assures seamless communications among first responders and emergency personnel at the scene of a major disaster. To address this problem, Congress included provisions in the Middle Class Tax Relief and Job Creation Act of 2012 (P.L. 112-96) for planning, building, and managing a new, nationwide, broadband network for public safety communications, and assigned additional spectrum to accommodate the new network. In addition, the act has designated federal appropriations of over $7 billion for the network and other public safety needs.

These funds will be provided through new revenue from the auction of spectrum licenses. The cost of construction of a nationwide network for public safety is estimated by experts to be in the tens of billions of dollars over the long term, with similarly large sums needed for maintenance and operation. In expectation that public-private partnerships to build the new network will reduce costs to the public sector, the law has provided requirements and guidelines for shared use.

The act has mandated that technical standards developed for the new network incorporate commercial standards for Long Term Evolution (LTE). LTE is a fourth-generation wireless technology that bases its operating standards on the Internet Protocol (IP). IP-enabled networks and wireless devices provide higher capacity and transmission speeds than earlier generations of technology. LTE represents the convergence of wireless technology with the Internet, bringing the capacity and resiliency of packet-switched networks to emergency communications. It is generally believed that the use of LTE and IP standards will greatly enhance communications for emergency response and recovery.

There are many challenges for public safety leaders and policy makers in establishing IP-enabled technologies as the baseline for the development of future solutions for response and recovery. One of the immediate challenges in developing standards is the need for a clear policy on the use of spectrum for commercial and public safety LTE. Because public safety planning has lagged behind commercial efforts to build LTE networks, the work on design and development of technical requirements is incomplete. Many experts are concerned that these delays may place public safety officials at a disadvantage in negotiating with potential partners, increase costs, and add further delays in moving forward to build a nationwide broadband network. Requirements in the act for standards development may be insufficient to overcome current technical obstacles for desired network features such as roaming between public safety and commercial networks.

In addition to monitoring progress in building the new broadband network for public safety, Congress may want to consider new policies for spectrum management and wireless innovation that would facilitate the transition to IP-enabled networks. Acceleration of innovation in next-generation wireless technologies would likely benefit not only public safety communications but also all consumers of wireless service and the American economy.

Chapter 2 – Emergency responders across the nation rely on land mobile radio (LMR) systems to gather and share information and coordinate their response efforts during emergencies. These public safety communication systems are fragmented across thousands of federal, state, and local jurisdictions and often lack "interoperability," or the ability to communicate across agencies and jurisdictions. To supplement the LMR systems, in 2007, radio frequency spectrum was dedicated for a nationwide public safety broadband network. Presently, 22 jurisdictions around the nation have obtained permission to build public safety broadband networks on the original spectrum assigned for broadband use. This requested report examines (1) the investments in and capabilities of LMR systems; (2) plans for a public safety broadband network and its expected capabilities and limitations; (3) challenges to building this network; and (4) factors that affect the prices of handheld LMR devices. GAO conducted a literature review, visited jurisdictions building broadband networks, and interviewed federal, industry, and public safety stakeholders, as well as academics and experts.

In: Public Safety Broadband ...
Editor: Thomas Giordano

ISBN: 978-1-62417-524-4

Chapter 1

The First Responder Network and Next-Generation Communications for Public Safety: Issues for Congress*

Linda K. Moore

Summary

Since September 11, 2001, when communications failures contributed to the tragedies of the day, Congress has passed several laws intended to create a nationwide emergency communications capability. Yet the United States has continued to strive for a solution that assures seamless communications among first responders and emergency personnel at the scene of a major disaster. To address this problem, Congress included provisions in the Middle Class Tax Relief and Job Creation Act of 2012 (P.L. 112-96) for planning, building, and managing a new, nationwide, broadband network for public safety communications, and assigned additional spectrum to accommodate the new network. In addition, the act has designated federal appropriations of over $7 billion for the network and other public safety needs. These funds will be provided through new revenue from the auction of spectrum licenses. The cost of construction of a nationwide network for public safety is estimated by

* This is an edited, reformatted and augmented version of the Congressional Research Service Publication, CRS Report for Congress R42543, dated August 29, 2012.

experts to be in the tens of billions of dollars over the long term, with similarly large sums needed for maintenance and operation. In expectation that public-private partnerships to build the new network will reduce costs to the public sector, the law has provided requirements and guidelines for shared use.

The act has mandated that technical standards developed for the new network incorporate commercial standards for Long Term Evolution (LTE). LTE is a fourth-generation wireless technology that bases its operating standards on the Internet Protocol (IP). IP-enabled networks and wireless devices provide higher capacity and transmission speeds than earlier generations of technology. LTE represents the convergence of wireless technology with the Internet, bringing the capacity and resiliency of packet-switched networks to emergency communications. It is generally believed that the use of LTE and IP standards will greatly enhance communications for emergency response and recovery.

There are many challenges for public safety leaders and policy makers in establishing IP-enabled technologies as the baseline for the development of future solutions for response and recovery. One of the immediate challenges in developing standards is the need for a clear policy on the use of spectrum for commercial and public safety LTE. Because public safety planning has lagged behind commercial efforts to build LTE networks, the work on design and development of technical requirements is incomplete. Many experts are concerned that these delays may place public safety officials at a disadvantage in negotiating with potential partners, increase costs, and add further delays in moving forward to build a nationwide broadband network. Requirements in the act for standards development may be insufficient to overcome current technical obstacles for desired network features such as roaming between public safety and commercial networks.

In addition to monitoring progress in building the new broadband network for public safety, Congress may want to consider new policies for spectrum management and wireless innovation that would facilitate the transition to IP-enabled networks. Acceleration of innovation in next-generation wireless technologies would likely benefit not only public safety communications but also all consumers of wireless service and the American economy.

INTRODUCTION

The importance of wireless communications in emergency response has expanded in parallel with increasing reliance on mobile communications across all sectors of the American economy. The consequences of failure in emergency communications networks have also grown, as the nation witnessed on September 11, 2001, and in the days that followed, as first responders and other emergency workers struggled to communicate with each other. The need for robust emergency communications was again underlined by network failures in the wakes of Hurricanes Katrina and Rita, in 2005. Fixing the problems of communications interoperability and operability that hampered response and recovery in these and other catastrophic events has been and remains a long-term goal of policy makers.[1]

After September 11, many experts recognized that a first responder communications network with national coverage would provide the standards and connectivity needed for interoperability and survivability. The National Commission on Terrorist Attacks Upon the United States (9/11 Commission) also recognized the role of networks in providing interoperability, citing the Army Signal Corps as a possible model in recommendations to Congress.[2]

From 2002 through 2007 Congress passed several laws intended to provide the Department of Homeland Security with the tools to plan for a national network. Efforts fell short of congressional expectations, however, in part because federal resources were directed to maintaining local jurisdiction in decision-making at the expense of coordinating a nationwide network.[3]

With the passage of the Middle Class Tax Relief and Job Creation Act of 2012 (P.L. 112-96) on February 22, 2012, the Administration, Congress, the public safety sector, and many other stakeholders have come together to begin the process of developing, constructing, and operating a nationwide network designed to meet public safety communications needs. The act has given government agencies and public safety officials new tools for providing nationwide availability of state-of-the art communications capability for emergency response and recovery. A new network is to be built to provide broadband communications using Internet Protocol (IP) standards to support high-speed delivery of data-rich content and video (broadband). Mission critical voice communications using standards designed for Land Mobile Radio (LMR) will be carried on separate networks. In time, many anticipate that IP standards for radios will replace LMR, bringing new economies of scale and higher levels of performance. The development of a unified effort to provide a national network places the nation on the path to achieve the long-sought goal of robust, interoperable communications for first responders.

The IP-based technology mandated for the nationwide public safety broadband network is Long Term Evolution (LTE). LTE networks and wireless devices are already being deployed in the commercial sector.

KEY NEW LEGISLATIVE PROVISIONS TO IMPROVE PUBLIC SAFETY COMMUNICATIONS

A program to provide nationwide coverage for public safety communications is to be developed and managed by a new authority created in Title VI of the Middle Class Tax Relief and Job Creation Act of 2012 (P.L. 112-96). The First Responder Network Authority, or FirstNet, has been established by the act and given broad powers to ensure that the nationwide public safety broadband network is built, maintained, and kept up-to-date as technology evolves.[4] In consultation with federal and state authorities, FirstNet will develop proposals to construct and manage the network with partners from the private sector, among others. Following is a discussion of major provisions in the act that pertain to public safety communications, including provisions to improve the nation's 911 emergency call system.

Among federal agencies designated by the act to provide consultation and support are the Federal Communications Commission (FCC), the National Telecommunications and Information Administration (NTIA), the National Institute of Standards and Technology (NIST), and the Office of Emergency Communications (OEC). The FCC manages commercial and non-federal spectrum use, including spectrum allocated to public safety. The NTIA manages federal spectrum resources and, along with NIST, is an agency within the Department of Commerce. OEC is part of the Office of Cybersecurity and Communications, Department of Homeland Security.

Spectrum Assignment

Radio frequency spectrum is an essential resource for wireless communications. The energy in electronic telecommunications transmissions converts airwaves into channels to deliver voice, text, and images. These channels are allocated for specific purposes, such as television broadcasting or WiFi,[5] and assigned to specific users through licenses. Allocating sufficient spectrum for wireless emergency communications has long been a concern for Congress. The Balanced Budget Act of 1997 (P.L. 105-33), for example, directed the FCC to allocate 24 MHz[6] of spectrum in the 700 MHz band for public safety use.[7]

With the passage of the Middle Class Tax Relief and Job Creation Act of 2012, some existing public safety licenses in the 700 MHz band[8] and an additional license (known as the D Block),[9] together totaling 22 MHz, have been designated by Congress to support a broadband communications network for public safety. The initial, 10-year license is required to be assigned by the FCC to FirstNet. It is renewable for an additional 10 years, on condition that FirstNet has met its duties and obligations under the act.[10]

A total of 34 MHz of spectrum capacity will therefore be available for public safety networks within the 700 MHz band: the 22 MHz designated for broadband and 12 MHz allocated for narrowband communications, primarily voice.[11] Additionally, there are public safety networks on adjacent frequencies within the 800 MHz band. Time and technological advances may someday bring these spectrum assets together, but at present there are three distinct public safety network technologies in use or planned within the 700 MHz and 800 MHz bands. These are: broadband communications at 700 MHz; interoperable narrowband communications at 700 MHz; and narrowband communications at 800 MHz. Some of the narrowband networks at 700 MHz and 800 MHz can share infrastructure and radios but older narrowband networks at 800 MHz are often not easily integrated with narrowband networks being built on 700 MHz frequencies.

All of the 700 MHz band spectrum assigned for public safety use can support broadband networks. At present, however, there is no tested technology to deliver voice communications over LTE broadband that meets first responder needs. The act gives the FCC the authority to "... allow the narrowband spectrum to be used in a flexible manner, including usage for public safety broadband communications...." subject to technical and interference protection measures.[12] This provision might open an opportunity for public safety agencies that want to be in the vanguard of using voice communications technology on LTE networks.

To offset the reassignment of the D Block from commercial to public safety use, the act requires the eventual return of frequencies known as the T-Band.[13] These are frequencies between 470 and 512 MHz allocated for television that have been made available for public safety use in 11 urban areas.[14] Since the transition to digital television, radio transmissions on some of these frequency assignments have experienced interference and the public safety agencies that use them are considering moving to new networks at 700 MHz. Other areas have recently invested to upgrade networks built on the T-Band frequencies and are concerned about the loss of this communications capacity. The act requires that the FCC act by February 2021 to establish a relocation plan that would free up the T-Band for reassignment through competitive bidding. Proceeds from the auctions of T-Band frequencies are to be available for grants to cover relocation costs. There are no requirements in the law as to how the NTIA, the designated grants administrator, is to structure the grant program or determine eligible costs, although the agency might decide to follow procedures for reallocating federal spectrum. The FCC and the NTIA might choose to work together to develop a transition plan that would allow public safety agencies that want to reallocate to 700 MHz to do so on an expedited basis.

Some of the earliest spectrum assignments for public safety are in channels below 512 MHz. Public safety and other license-holders in designated channels below 512 MHz are required to reband their holdings to conform to an FCC mandate to improve spectrum efficiency.[15] This narrowbanding requirement, as it is called, requires that assigned channels be reduced from a width of 25 khz to 12.5 khz, thereby freeing up new spectrum capacity for public safety and other uses. The deadline to meet the narrowbanding requirement is January 1, 2013. To accommodate public safety licenseholders in the T-Band that now fall under requirements established in the act, the FCC has ruled to exempt them from the narrowbanding requirements.[16]

Other spectrum assets available for public safety communications include 50 MHz of spectrum at 4940-4990 MHz (4.9 GHz). Current technology limits these frequencies to local area networks covering a small area. Many experts believe that short-range applications can be incorporated with broadband networks to provide additional resources and better coverage, for example in responding to emergencies in high-rise buildings.[17]

To maximize the effectiveness of these multiple spectrum holdings, many believe that additional technological development and planning should be undertaken. FirstNet's mandate appears to limit it to the public safety broadband network to be operated on the spectrum licensed to it. Although not specifically required by the act, several federal agencies do have broad powers to undertake research and development that might further goals for improved performance of emergency communications systems, and more efficient and effective use of all spectrum resources allocated for public safety use.

Expenditures and Revenue Sources

The cost of building a new wireless communications network is likely to be in the tens of billions of dollars.[18] The expectation is that FirstNet will have access to existing infrastructure for some of the network's components and that it will be able to invest through partnerships—with commercial wireless carriers or other secondary users of its spectrum and infrastructure—that generate revenue.

The Middle Class Tax Relief and Job Creation Act of 2012 provides over $7 billion in funding either to FirstNet and states participating in the nationwide network, or as grants to states that have opted out of participating in the FirstNet nationwide network program, but have qualified to build their state's portion of the nationwide network. There is an initial loan of $2 billion (repayable from spectrum-license auction proceeds) to set up FirstNet and begin its operation.[19] The remaining $5 billion will become available as auctions for spectrum licenses are concluded and the revenues deposited in the Public Safety Trust Fund.

Public Safety Trust Fund

The law provides for transfers from a Public Safety Trust Fund that is created by the act to receive revenues from designated auctions of spectrum licenses.[20] The designated amounts are to remain available through FY2022, after which any remaining funds are to revert to the Treasury, to be used for deficit reduction. Auction proceeds are to be distributed in the following priority:

- To the NTIA, to reimburse the Treasury for funds advanced to cover the initial costs of establishing FirstNet: not to exceed $2 billion.
- To the State and Local Implementation Fund for a grant program: $135 million.
- To the Network Construction Fund for costs associated with building the nationwide network and for grants to states that qualify to build their own networks: $7 billion, reduced by the amount advanced to establish FirstNet.
- To NIST for public safety research: $100 million.
- To the Treasury for deficit reduction: $20.4 billion.
- To the NTIA and the National Highway Traffic Safety Administration for a grant program to improve 911 services: $115 million.
- To NIST for public safety research: $200 million.
- To the Treasury for deficit reduction: any remaining amounts from designated auction revenues.

Network Construction Fund

The Network Construction Fund is established in the Treasury to be used by FirstNet for expenditures on construction, maintenance, and related expenses to build the nationwide network required in the act, and by the NTIA for grants to those states that qualify to build their own radio access network links to the FirstNet core infrastructure.[21]

FirstNet: Limit on Expenditures

The act caps FirstNet's administrative expenses at $100 million in total over the first 10 years of operation. Costs attributed to oversight and audits are not included in the expense cap.[22]

FirstNet: Fee Income and Other Revenue

FirstNet has the authority to obtain grants and to receive payment for the use of network capacity licensed to FirstNet and of network infrastructure "constructed, owned, or operated" by FirstNet.[23] Specifically, FirstNet is authorized to collect network user fees from public safety and secondary users[24] and to receive payments under leasing agreements in public-private partnerships.[25] These partnerships may be formed between FirstNet and a secondary user for the purpose of constructing, managing, and operating the network. The agreements may allow access to the network on a secondary basis for services other than public safety. FirstNet and its partners may also receive payments for leasing access to infrastructure, such as towers.[26] The act requires that these fees be sufficient each year to cover annual expenses of FirstNet to carry out required activities,[27] with any remaining revenue going to network construction, operation, maintenance, and improvement.[28] There is a prohibition on providing service directly to consumers; this does not impact the right to collect fees from a secondary user or enter into leasing agreements.[29]

State and Local Implementation Fund

The State and Local Implementation Fund is to receive $135 million from the Public Safety Trust Fund. The NTIA, which administers the grant program for this fund, may borrow up to the full amount.[30] The grants are to go to states to develop a plan on how to use a nationwide public safety broadband network to meet their emergency communications needs. The program is to be established as a matching grant program. Federal grants from the fund are not to exceed 80% of the projected cost to the state, however, the NTIA may make the decision to waive the matching funds requirement.[31] The distribution of available funds among the states will be established by the NTIA in consultation with FirstNet.[32]

In compliance with the act's deadline for setting up the State and Local Implementation Fund, the NTIA has published initial programmatic requirements under which it will award grants.[33] The notice states that the NTIA expects to consult with FirstNet on the scope of the grants with the goal of issuing grant requirements in the first quarter of 2013. Funding is planned for distribution in two phases. The first phase will provide funding for initial planning and related activities. The second phase will begin after FirstNet has consulted with the 56 states and territories eligible to receive grants under the program. The second phase will address states' needs in preparing for additional consulting with FirstNet, and for planning to undertake data collection activities. Among the requirements that the NTIA is considering for grant eligibility are requirements for states to identify potential public safety users of the network; to develop standard Memoranda of Understanding to facilitate the use of existing infrastructure; and to prepare a comprehensive plan "describing the public safety needs that they expect FirstNet to address in its design" of the network.[34]

Other Sources of Funds

The construction of this new network represents a significant investment for all participants. State public safety agencies have multiple obligations to build or upgrade, and equip, other networks and may not be in a position to contribute to building and maintaining the new broadband network. The ability of FirstNet to procure funding from the private sector may be crucial to its success.

Planning Authority

The Middle Class Tax Relief and Job Creation Act of 2012 creates FirstNet as an independent entity within the NTIA and empowers it to oversee the establishment of an interoperable broadband network for public safety. The act requires that state and local agencies have a consulting role in the development, deployment, and operation of the nationwide network. The act further provides an opportunity for states to build their own radio access networks within the framework of the nationwide broadband network.

FirstNet

FirstNet is to be headed by a board of 15 members of which 12 are to be appointed by the Secretary of Commerce according to criteria established by the law, which are intended to provide both representation from key stakeholders and expertise. The other three members of the board are the Secretary of the Department of Homeland Security, the Attorney General of the United States, and the Director of the Office of Management and Budget. The Secretary of Commerce is required to appoint a chairman of the board for an initial term of two years.[35] FirstNet has the statutory authority to exercise all powers specifically granted by the act and "such incidental powers as shall be necessary."[36] Appointments to the board were announced on August 20, 2012.[37] One of the board's first steps is likely to be the appointment of the public safety advisory council required by the act.

FirstNet is required to create a public safety advisory committee to assist in carrying out its mandate.[38] The committee is to take "all actions necessary to ensure the building, deployment, and operation" of the network in consultation with federal, state, tribal, and local public safety entities, the Director of NIST, the FCC, and the public safety advisory committee.[39] There are no requirements in the statute as to the composition of the committee.

FirstNet appears to be an autonomous organization, with broad powers to carry out its mandate, within the requirements established by the law. It has for example sole power to select the program's manager and its agents, consultants, and other experts subject to the requirement that they be chosen "in a fair, transparent, and objective manner."[40] In managing proposals and contracts, it is to "take such other actions as may be necessary" to accomplish the network buildout.[41]

As part of its management of the network, FirstNet is required at a minimum:

- To establish network policies, including development of detailed requests for proposals to build the network, and operational matters such as terms of service and billing practices.[42]
- To consult with states on expenditures, as part of the preparation of policies and requests for proposals.[43]
- To enter into agreements to use existing communications infrastructure, including commercial and federal infrastructure, "to the maximum extent economically desirable."[44]

- To ensure the construction, maintenance, operation, and improvement of the broadband network, taking into account new and evolving technologies.[45]
- To enter into agreements with commercial networks to allow public safety roaming on their networks.[46]
- To represent the interests of the network's users before standards-setting boards, in consultation with NIST, the FCC, and its own Public Safety Advisory Committee.[47]

State and Local Participation

Every state has one or more agencies that plan for public safety, homeland security, and emergency communications. Most states have a Statewide Interoperability Coordinator (SWIC) to administer its Statewide Communication Interoperability Plan (SCIP).[48] SCIPs are written to conform with federal guidelines and requirements, such as the National Emergency Communications Plan. FirstNet is required to consult with regional, state, tribal, and local authorities regarding decisions such as those concerning the costs of the policies it formulates, as required in the law, including expenditures for the core network, placement of towers, coverage areas, security, and priority access for local users. Consultation will be through a state-selected coordinator as specified in the act.[49] Appointment of an individual or governmental body as the point-of-contact is also required as a condition of state participation and eligibility to receive grants established by the act.[50] States may decide to use the existing SWIC as the required single point-of-contact or may choose to appoint a separate coordinator.

The governor of each state is to be notified by FirstNet when it has completed its requests for proposals regarding construction, operation, maintenance, and improvement of a nationwide network. The governor or his designee will receive the details of the proposed plans and notification of the amount of funding available to the state if it participates in the FirstNet program.[51]

A state that does not want to participate in FirstNet must submit an alternative plan for construction, operation, maintenance, and improvement of the radio access network within the state. The state must demonstrate to the FCC, which the law requires to review the plan, that its planned network would comply with minimum technical requirements and be interoperable with FirstNet. The state has 90 days to agree to participate or to notify

FirstNet, the NTIA, and the FCC of its intent to deploy its own part of the radio access network, and an additional 180 days to provide its plan to the FCC.[52]

If the FCC does not approve the plan, the state might be obliged to participate in FirstNet.[53] If a state's plan is approved it will be eligible to apply for a grant, administered by the NTIA, that will be funded from the Network Construction Fund created by the act. The amount available will be less than what would have been provided if the state had opted in to the FirstNet program, because the grant will be applied only toward building the radio access network and may be subject to matching grant requirements. Approval of the grant is contingent on meeting additional requirements established by the NTIA, including sustainability, timeliness, cost-effectiveness, security, coverage, and services that are comparable to FirstNet.[54] The state would be required to pay a user fee for access to FirstNet.[55] It would not be permitted to enter commercial markets or lease access to its network except through a public-private partnership. Any revenue to the state from a partnership must be used only for costs associated with its broadband network.[56]

Some industry observers have expressed concern about the impact on the success of the nationwide broadband network if many states choose to build their own radio access networks. The cost to FirstNet of building the nationwide network may go up, for example, if anticipated economies of scale are diminished. It may be more difficult for FirstNet to negotiate the partnerships that are expected to provide much of the needed funding for the network. A state that has its plans approved by the FCC may not be able to meet stipulated requirements when its network is built; absent any action by the FCC to enforce technical requirements, the goal of seamless interoperability across all broadband systems may be jeopardized. States may also have difficulty in finding the funds to complete radio access network build-outs, leaving significant gaps in what is intended to be nationwide coverage. The law only identifies two options for a state: join FirstNet or build a statewide radio access network subject to the provisions of the act. It does not say whether states may choose to opt-out of the broadband network entirely, choosing neither to join FirstNet nor to build a broadband network on the frequencies assigned to FirstNet. Some states may prefer to concentrate their resources on improving mission-critical voice networks and acquire broadband access from a commercial provider or through other means.

One advantage for states building their own radio access networks on FirstNet spectrum is that they will have greater control over any partnerships formed and on expenditures within their states. Although the act requires states to use any revenue from partnerships only to cover costs associated with the state's network, the states will be able to make their own decisions about priorities, with more confidence that revenues will be available when needed. Although there are many potential benefits for states to participate in a nationwide network, such as economies of scale, more secure and robust communications, and a unified base for collaborative efforts, there are also a number of risks, especially if FirstNet fails to deliver the benefits. The success of FirstNet as an accepted planning authority and leader may depend on whether it makes a compelling business case in the requests for proposals required by the act.

FirstNet's plans for partnerships with the private sector and the nature of the network development plans proposed to each state may be of particular interest to Congress as an early indicator of the viability of FirstNet in meeting the goals required by the act.

Federal Governance

Federal governance of the nationwide public safety broadband network, as required by the Middle Class Tax Relief and Job Creation Act of 2012, is primarily through consultation and oversight. Planning, investment, operating, and other related decisions are to be made by the FirstNet board and the experts it is to hire on a permanent or consultative basis. The designated appropriate congressional committees are, in the Senate, the Committee on Commerce, Science, and Transportation; in the House, the Committee on Energy and Commerce.[57] These committees and other committees with jurisdiction are likely to take an active role in oversight, many believe.

Examples of statutory obligations for Congress and the Administration in the direction of FirstNet include:

Membership on FirstNet board. The members of the FirstNet board are to be chosen by the Secretary of Commerce, within the parameters established in the act. The Department of Homeland Security, the Attorney General, and the Office of Management and Budget each have one member on the board in permanence. The Secretary of Commerce is required to appoint a chairman of the board for an initial term of two years.[58]

Grant programs for planning. The NTIA is to establish and administer the State and Local Implementation Fund. Grant provisions are to be in accordance with decisions made by FirstNet.[59]

Grant programs for state networks. The NTIA is to administer grants from the Network Construction Fund to states that qualify to build their own radio access networks and choose to apply for a grant.[60]

Spectrum leases for state networks. The NTIA sets the terms and is responsible for enforcing the requirement that states qualifying to build their radio access networks must sublease spectrum through FirstNet, the assigned license-holder.[61]

License review. The FCC is required to review the initial 10-year license assigned to FirstNet and consider its renewal based on performance criteria.[62]

Performance review. The Government Accountability Office (GAO), within 10 years, is to prepare a report assessing the effectiveness of FirstNet with recommendations on "what action Congress should take" regarding the mandated termination of authority.[63]

Fee schedule. The NTIA is to review and approve the annual schedule of fees charged to public safety agencies and other users for access to FirstNet's resources.[64]

Annual audit. The Secretary of Commerce is to contract for an annual audit of FirstNet's finances and activities. The reports are to be submitted to Congress, the President, and FirstNet.[65]

Report to Congress. FirstNet is required to submit annual reports to Congress on its "operations, activities, financial conditions, and accomplishments."[66]

Although there are several platforms for oversight and guidance provided in the act, it seems likely that the primary responsibility for monitoring progress will fall to the NTIA. The agency may choose to seek assistance from other agencies beyond what is specified in the act, possibly through memoranda of understanding.

Public-Private Partnerships

Partnerships are expected to play a critical role in building and operating the network. Electric utility companies, for example, are upgrading their networks to meet Smart Grid requirements,[67] and some companies have expressed an interest in partnering with FirstNet or state authorities. Some commercial wireless service providers have also expressed an interest in working in partnership with public safety entities to develop and operate new broadband networks.

The Middle Class Tax Relief and Job Creation Act of 2012 requires FirstNet to issue "open, transparent, and competitive" requests for proposals to private sector entities for building, operating, and maintaining the network[68] that leverage to the extent "economically desirable" existing commercial wireless infrastructure, in order to expedite network deployment.[69] It is charged with managing and overseeing the resulting contracts or agreements. As part of a separate requirement to assure substantial rural coverage during all phases of deployment, the act requires that industry proposals and contracts include, if possible, partnerships with existing commercial mobile providers.[70]

Decisions by FirstNet about the network's design, construction, and operation are likely to have a significant impact on commercial participation in a public safety broadband network or networks. These decisions may also influence decision-making by states as to whether or not to pursue radio area network construction independently or through their own partnerships.

Congress may be interested in the composition of private sector partnerships formed by FirstNet and individual states, not only for their business plans but also for the inclusion of a wide variety of stakeholders. For example, are rural and tribal wireless carriers included as business partners? Do secondary access agreements support services that meet social goals, such as for telemedicine, or are they exclusively for commercial purposes? Is competition in providing wireless services being enhanced or hindered?

Infrastructure

Infrastructure for the new network includes operations centers, towers, antennae, and other communications equipment, as well as radios and the software that links them to the network. For wireless communications, an important infrastructure component is the network that links radio towers to communications backbones. These networks, which usually operate over fiber-optic cable or microwave connection, are typically referred to as backhaul.

The Middle Class Tax Relief and Job Creation Act of 2012 requires FirstNet to establish a nationwide, interoperable public safety network,[71] with a "single, national network architecture that evolves with technological advancement.... "[72] Network infrastructure components that are specifically required include:

- Core network of national and regional data centers and other elements, all based on commercial standards.
- Connectivity between the radio access network and the public Internet or the Public Switched Telephone Network, or both.
- Network cell site equipment, antennas, and backhaul equipment, based on commercial standards, to support wireless devices operating on frequencies designated for public safety broadband.

FirstNet is required to leverage existing infrastructure by entering into agreements to use commercial or other communications infrastructure including federal, state, tribal, or local infrastructure.[73] Planned phases for infrastructure deployment are to include "substantial rural coverage."[74]

FirstNet's ability to build the required network may depend on the timeliness, scope, and outcome of its negotiations to share infrastructure with other parties in order to focus resources on providing elements deemed essential for public safety use of broadband communications.

Timeframe

The requirements of the Middle Class Tax Relief and Job Creation Act of 2012 must be substantially met and the viability of the project demonstrated no later than the end of FY2022, if not sooner. The State and Local Implementation Fund and the Network Construction Fund expire in 2022, with any balances reverting to the Treasury. By 2022, the GAO must have begun an assessment of the performance of FirstNet and the FCC must decide whether or not to renew the licenses for the public safety broadband network. Within this 10-year timeframe, there are few deadlines beyond requirements for the initial establishment of the planning and implementation framework.

Deadlines for the FCC required the prompt preparation of recommendations for minimum technical requirements for interoperability to be presented to FirstNet.[75] The Secretary of Commerce was given a somewhat longer deadline to appoint the board members of FirstNet (180 days from enactment).[76] Within the same timeframe, the NTIA is to establish grant program requirements for the State and Local Implementation Fund, in consultation with the FirstNet board.[77] By the end of August 2012, therefore, key components of the program for a new communications network should be in place.

Many of the next important steps for building the network have no required deadline. Some milestones, such as rural coverage, are mandated in the act, but the deadlines are not specified. There are, for example, no deadlines in provisions that require FirstNet to:

- Establish a standing committee on public safety.[78]
- Develop requests for proposals that include a requirement for timetables; and to consult with states on establishing state and local planning processes.[79]
- Complete the request for proposal process that is to be given to each state governor regarding the request for proposal and its details, and the funding level for each state as determined by the NTIA.[80]

Mandated deadlines for states include

- Within 90 days of receipt of notice from FirstNet, the governor shall choose either to participate in deployment of FirstNet or to conduct its own deployment within the state.[81]
- Within 180 days of giving notice to opt out of FirstNet, the governor shall complete requests for proposals for a state network.[82]

No deadline is established in the statute for the FCC to approve or disapprove state proposals for their own portion of the nationwide broadband network.[83] There are also no specified deadlines for a state to apply to the NTIA for a grant to construct the radio access network and to lease spectrum capacity from First Net, if FCC approval is received for a state network.[84] However, one condition of eligibility for a grant to a state to build its own radio access network is that the state's plan must demonstrate "the ability to complete the project within specified comparable deadlines...."[85]

FirstNet and the FCC may need to be expeditious in completing all steps for the preparation, review, and acceptance of requests for proposals so that construction of the required core network begins in a timely manner. Too many delays in administrative processes may erode the feasibility of the project. An official of the NTIA has reportedly predicted that it will take at least a year for the initial planning and preparatory work to be completed before network construction might begin.[86]

Next Generation 9-1-1

Today's 911 system is built on an infrastructure of analog technology that does not support many of the features that most Americans expect to be part of an emergency response. Efforts to splice newer, digital technologies onto this aging infrastructure have created points of failure where a call can be dropped or misdirected, sometimes with tragic consequences. Callers to 911, however, generally assume that the newer technologies they are using to place a call are matched by the same level of technology at the 911 call centers, known as Public Safety Answering Points (PSAPs). However, this is not always the case. To modernize the system to provide the quality of service that approaches the expectations of its users will require that the PSAPs and state, local, and possibly federal emergency communications authorities invest in new technologies. As envisioned by most stakeholders, these new technologies—collectively referred to as Next Generation 911 or NG9-1-1—should incorporate Internet Protocol standards. An IP-enabled emergency communications network that supports 911 will facilitate interoperability and system resilience; improve connections between 911 call centers; provide more robust capacity; and offer flexibility in receiving and managing calls. The same network can also serve wireless broadband communications for public safety and other emergency personnel, as well as other purposes.

Recognizing the importance of providing effective 911 service, Congress has passed three major bills supporting improvements in the handling of 911 emergency calls. The Wireless Communications and Public Safety Act of 1999 (P.L. 106-81) established 911 as the number to call for emergencies and gave the Federal Communications Commission (FCC) authority to regulate many aspects of the service. The most recent of these laws, the NET 911 Improvement Act of 2008 (P.L. 110-283), required the preparation of a National Plan for migrating to an IP-enabled emergency network. Responsibility for the plan was assigned to the E-911 Implementation Coordination Office (ICO), created to meet requirements of an earlier law, the ENHANCE 911 Act of 2004 (P.L. 108-494). Authorization for the ICO terminated on September 30, 2009. ICO was jointly administered by the

National Telecommunications and Information Administration and the National Highway Traffic Safety Administration.

The Middle Class Tax Relief and Job Creation Act of 2012 re-establishes the federal 9-1-1 Implementation Coordination Office (ICO) to advance planning for next-generation systems and to administer a grant program.[87] ICO is to provide matching grants to eligible state or local governments or tribal organizations for the implementation, operation, and migration of various types of 911 and IP-enabled emergency services, and for public safety personnel training.[88] States that have diverted fees collected for 911 services are not eligible for grants under the program.[89] Based on the act's prioritized plan for funding programs with spectrum license auction revenue, the funds for the grant program will be made available only after $27.635 billion of available auction revenue has been applied to other purposes.

Provisions in the act regarding 911 programs include:

- The GAO is required to study how states assess fees on 911 services and how those fees are used.[90]
- The General Services Administration is required to prepare a report on 911 capabilities of multi-line telephone systems in federal facilities and the FCC is to seek comment on the feasibility of improving 911 identification for calls placed through multi-line telephone systems.[91]
- The FCC is to assess the legal and regulatory environment for development of NG9-1-1 and barriers to that development, including state regulatory roadblocks.[92] The FCC is also to (1) initiate a proceeding to create a specialized Do-Not-Call registry for public safety answering points, and (2) to establish penalties and fines for autodialing (robocalls) and related violations.[93]
- ICO, in consultation with NHTSA and DHS is to report on costs for requirements and specifications of NG9-1-1 services, including an analysis of costs, and assessments and analyses of technical uses.[94]
- Immunity and liability protections are provided—to the extent consistent with specified provisions of the Wireless Communications and Public Safety Act of 1999—for various users and providers of Next Generation 911 and related services, including for the release of subscriber information.[95]

The act also requires FirstNet to promote integration of the nationwide public safety broadband network with PSAPs.[96] Since the NTIA has responsibilities for both ICO and FirstNet, the agency is in a position to improve interoperability between PSAPs and First Responders as they move to common IP-based platforms.

Technology and Standards

Standardization of network components, including radios, is generally considered essential to achieving interoperability, improving service, and reducing operating costs. The mandated standard for the new public safety network is Long Term Evolution (LTE), with technical requirements based on commercial standards for LTE.[97] LTE is a fourth-generation (4G) technology based on the Internet Protocol. The commercial sector has begun the transition to operating on IP-enabled networks such as LTE. Wireless carriers around the world are installing LTE networks for consumers and planning for the next generation of LTE: LTE Advanced.[98] LTE Advanced technologies will be able to operate across noncontiguous spectrum bands, thereby increasing channel widths for greater capacity and performance. Most experts agree that LTE Advanced will facilitate the transition to new technologies by making it easier and less expensive to phase out older infrastructure.

FirstNet

The Middle Class Tax Relief and Job Creation Act of 2012 requires FirstNet to assure nationwide standards for use of and access to the network it is tasked with developing. The act specifies the use of commercial standards for some of the network components.[99] In consultation with NIST, the FCC, and the Public Safety Advisory Committee, FirstNet is to represent the networks and its users before standards-setting organizations regarding standards related to interoperability.[100]

To promote competition, devices for public safety network radios and other wireless devices are required to be built to open, non-proprietary, commercially available standards, “capable of being used by any public safety entity and by multiple vendors across all broadband networks operating in the 700 MHz band” and backward compatible with existing commercial networks where necessary and feasible.[101]

FCC

The act requires the FCC to establish a Technical Advisory Board for First Responder Interoperability, and sets out criteria for the selection and participation of board members.[102] The primary purpose of the Interoperability

Board, as it is called, is to agree on minimum technical requirements for nationwide interoperability on the public safety broadband network. The Interoperability Board is required to develop these technical recommendations in consultation with the NTIA, NIST, and the OEC.[103] The board's technical recommendations are to be based on commercial standards for LTE.[104] The recommendations are to be delivered to the FCC by the end of May 2012.[105] The FCC then has 30 days to approve, modify if necessary, and deliver the recommendations to FirstNet.[106] The Interoperability Board is to be disbanded after the recommendations have been transmitted.[107]

The establishment of minimum technical requirements has a two-fold purpose. One, the requirements are to be presented to the Board of Directors of FirstNet as recommended requirements for interoperability. Second, the minimum technical requirements are to be used by the FCC as a standard of interoperability for evaluating state plans in cases where states have asked to build their own radio access networks.

In the report it submitted,[108] the Interoperability Board, in addition to minimum technical standards, also provided additional considerations that it judged to be important for achieving interoperability.

NIST

The Director of NIST, in consultation with the FCC, DHS, and the National Institute of Justice, Department of Justice, is to "conduct research and assist with the development of standards, technologies and applications to advance wireless public safety communications."[109] More specifically, in consultation with FirstNet and the Public Safety Advisory Committee, NIST is to

- Document technical requirements for public safety wireless communications.
- Accelerate the development of interoperability between currently deployed systems and the public safety broadband network.
- Establish a research plan and direct research for next-generation wireless public safety needs.
- Accelerate the development of broadband network features such as mission-critical voice, prioritization, and authentication.
- Accelerate the development of communications equipment and technology to facilitate the eventual migration of public safety narrowband communications to the public safety broadband network.[110]

Furthermore, the Director of NIST, in consultation with FirstNet and the FCC, "shall ensure the development of a list of certified devices and components meeting appropriate protocols and standards for public safety and commercial vendors" for those seeking to have the use of the public safety broadband network.[111]

OEC

The act's requirements for protecting and monitoring the network against cyberattack[112] are being addressed through the OEC, among others, in coordination with the NTIA. The OEC is leading a Cyber Risk Assessment for the nationwide public safety broadband network, analyzing the risk to cyber infrastructure in four parts: interoperability; operability; continuity; and security.[113]

Need for Standards Development

Narrowband and broadband networks for public safety will by most accounts be incompatible with each other and with other networks for the foreseeable future.[114] Only a small part of the existing public safety infrastructure is expected to be usable in the development of new networks at 700 MHz. To maximize the utility of new investments in infrastructure and radios, many believe that standards that support public safety applications for IP-enabled technologies must be completed in the early stages of planning and building. Just as access to the Internet has revolutionized business and social cultures worldwide, the transition to IP-enabled networks is likely to expand the capability and scope of emergency communications.

The act variously requires NIST, the FCC, and the NTIA[115] to develop standards and take steps to improve spectrum efficiency and support the development of the next generation of wireless technology. These agencies already have a number of initiatives in place, notably the Public Safety Communications Research program (PSCR). PSCR provides research, development, and testing to advance public safety communications interoperability. The program is a joint effort between NIST's Office of Law Enforcement Standards and NTIA's Institute for Telecommunication Sciences and is sponsored by the Office for Interoperability and Compatibility at DHS, and the Department of Justice Community Oriented Policing Services.[116]

The act also directs FirstNet to negotiate standards on behalf of public safety. To do this effectively, FirstNet may need a permanent place in the commercial forums where standards are negotiated and established. Such access may come through participation with private sector partners.

The funding for the federal research and development efforts described in the act is provided from spectrum license auction revenue. The timing of the auctions and the prioritization for distributing auction revenues are such that the funds designated for research and development may not be available for several years, if at all. Some of the act's provisions require the FCC to auction designated spectrum within three years.[117] Auction procedures require several steps that are published for comment before final rulemaking, and the process typically takes a year or more before an auction commences. The first round of funding for NIST ($100 million) would occur once the proceeds from spectrum license auctions deposited in the Public Safety Trust Fund surpass $7.135 billion. The second funding round for NIST would occur after deposits reach $27.75 billion. Although resources in existing federal programs may be shifted to give priority to the implementation of the Middle Class Tax Relief and Job Creation Act of 2012,[118] the federal government may not be able to fund all of the standards and other technological research that is required by the act or needed for public safety. Timely development of public safety applications for LTE and LTE Advanced may come primarily from the private sector, where some vendors are developing components needed for the broadband network and its devices. To meet its responsibilities under the act, FirstNet may choose to allocate some of the funding provided to it by the act, or raise additional funds, to facilitate standards development.

If no solution is found to coordinate private and public work on standards development and new technologies for emergency communications, the development of IP-enabled technologies for public safety may continue to lag behind that of the commercial sector, perpetuating the high costs and inefficiencies that have plagued first responder communications for decades.

INTEROPERABILITY WITHIN THE 700 MHZ BAND

In its *National Broadband Plan*, the FCC indicated that it wanted to make commercial networks in the 700 MHz band available for public safety use and requested that Congress confirm the FCC's authority to act.[119] The Middle Class Tax Relief and Job Creation Act of 2012 provides the FCC with statutory authority to establish rules in the public interest to improve the ability of public safety networks to roam on commercial space and to gain priority access.[120]

FirstNet and the states that build their own networks are empowered by the act to enter into agreements with commercial providers that would allow public safety network users to roam on partnering networks. Agreements might also cover rules for priority access in times of high demand for network capacity. Priority access can take several forms, such as "ruthless pre-emption," in which non-public-safety transmissions are immediately terminated to make way for emergency communications, or negotiated priority agreements that might, for example, place public safety users at the head of the line for network access as capacity becomes available. The act stipulates that the FCC's authority may not require roaming or priority access unless (1) the public safety and commercial networks are technically compatible; (2) the commercial network is reasonably compensated; and (3) access does not preempt or otherwise terminate or degrade existing traffic on the commercial network.[121] Within these limits, the FCC appears to have some leeway to use its regulatory authority to support public safety in negotiations with partners. The FCC cannot, under the act, mandate ruthless pre-emption, although the act does not preclude contractual negotiations that would allow it.

The act's provisions for roaming and priority access do not require a commercial vendor to make additional investments to insure technical compatibility, and the act's language might be interpreted as precluding an FCC mandate to that effect. Interpretation and enforcement of the compatibility provision may pose an obstacle to achieving desired levels of network interoperability and cross-network roaming because the current technical standards for the 700 MHz band preclude affordable full-spectrum roaming, that is, the ability of any network within the 700 MHz to roam on any other network within the 700 MHz band. Full-spectrum roaming is considered by many to provide advantages for public safety and also for the public at large. For example, it makes more network capacity available for shared emergency communications of all types, not just for first responders. Many believe that full-spectrum access supports competitiveness among wireless carriers—in particular assisting small wireless carriers serving rural areas to offer new broadband services—by providing access to all customers within the band.

Achieving full-spectrum roaming on the 700 MHz band requires modifications of technical requirements for LTE, the preferred technology for mobile broadband within the 700 MHz band. The standards for LTE are agreed on a global basis by the 3GPP, a standards setting group.[122] For a number of technical considerations, including maximizing spectrum efficiency and minimizing interference across spectrum channels, the 3GPP divided the

700 MHz commercial spectrum into different band classes. As documented by the FCC,[123] the 70 MHz of commercial spectrum within the 700 MHz band is the only non-interoperable commercial service band.

The band classes apply to spectrum blocks that were determined by the FCC. In preparing for an auction of spectrum licenses, the FCC follows a number of procedural steps and seeks comment on planned actions. A key first step is to develop a band plan for the spectrum and make decisions about geographical coverage and technical requirements. Allocation of spectrum blocks within the 700 MHz band occurred in several steps. To protect public safety allocations for narrowband networks in the upper end of the band from harmful interference, the FCC divided the band into an Upper 700 MHz Band and a Lower 700 MHz band. Both bands included paired spectrum licenses, that is, each license had two sets of channels, one for the uplink and the other for the downlink. The Lower 700 MHz Band conformed to industry standards for global cellular bands in the placement of the uplink and downlink; in the upper 700 MHz Band, the direction of the uplink and downlink were reversed. The auction of the majority of licenses for the 700 MHz band concluded on March 18, 2008. Standards for LTE (Release 8) were finalized in December of the same year.[124] A schematic of the auction blocks and band classes is provided in Figure 1.

The 700 MHz auction gave Verizon Wireless and AT&T Mobility dominant holdings in Band Class 17, corresponding to licenses in the B Block and C Block in the Lower 700 MHz Band, and in Band Class 13, corresponding to licenses in the C Block of the Upper 700 MHz Band. Many of the licenses in the A Block were acquired by smaller carriers. Because of restrictions on usage, licenses for frequencies in the A Block were less expensive than licenses in the B and C Blocks. The restrictions were largely, but not solely, to protect transmissions on these frequencies from potential interference from high-power TV broadcast signals located in TV channel 51, which is adjacent to the lower end of the 700 MHz band.

Public safety holdings include the D Block, which was originally designated for commercial use and assigned Band Class 14 by 3GPP. Band Class 14 was extended to include the part of the spectrum allocated to the nationwide public safety broadband network, as the FCC's band plan called for the two spectrum blocks to be shared. The narrowband public safety spectrum holdings at 700 MHz were not part of the 3GPP standards-setting process, as they are not commercial networks. Some licenses, such as the D and E Blocks in the Lower 700 MHz band, were auctioned before the development of LTE and have different technical requirements.

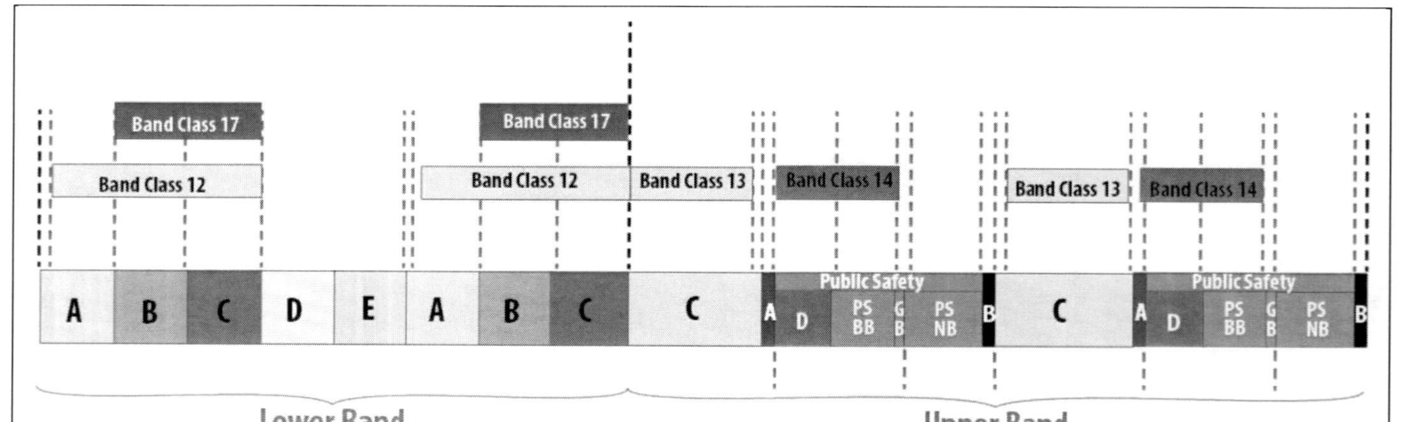

Source: FCC Notice of Proposed Rulemaking, "Interoperability of Mobile Use Equipment Across paired Commercial Spectrum Blocks in the 700 MHz Band," March 21, 2012.

Note: Within the public safety bands, spectrum allocated for broadband is BB, spectrum designated for narrowband is NB and guard bands to protect against interference are GB.

Figure 1. License Blocks and Band Classes in the 700 MHz Band.

As a consequence of incompatibility among the band classes, competition in the commercial sector and interoperability for public safety communications may be at risk. The difference between the standards for Band Class 12 and the other band classes has reportedly had the effect of isolating the A Block license-holders from the mainstream of development for LTE devices. Producers of the new equipment have tailored their first offerings for LTE deployment for the markets with the greatest demand: Band Classes 13 and 17. Smaller carriers have reported to the FCC that they are having difficulty in acquiring equipment, especially handsets, for Band Class 12, the only technical standard applicable for licenses in the A Block.

To address the concerns of carriers with licenses in the A Block, the FCC has opened a proposed rulemaking to address how to mitigate interference in Band Class 12 and to increase interoperability within the lower 700 MHz. The rulemaking does not specifically address the impact on public safety roaming and interoperability.

As is the case for Band Class 12, the costs of developing and producing the chipsets, software, and other components for equipment operating on Band Class 14 are likely to be spread across a relatively smaller customer base, increasing marginal costs and the prices paid by users. Because the band classes are not interoperable across the 700 MHz band, public safety network users are likely to incur not only higher costs for equipment to operate within their assigned frequencies, but also higher costs for roaming and priority access on commercial channels.

If FirstNet and any statewide network deployed in Band Class 14 choose different commercial partners, roaming and priority access may be limited. A network agreement, for example, with C Spire Wireless (formerly Cellular South) might lead to the development of equipment to operate on Band Classes 12 and 14; an agreement with Verizon, Band Classes 13 and 14; with AT&T, Band Classes 17 and 14. Some experts have raised concerns about the limitations that will be placed on public safety roaming as a result of the lack of interoperability within the 700 MHz band. The cost of public safety radios that can operate on all band classes will be high, if they can be engineered at all. A radio that only works on Band Class 14 and one other band class may be cut off from roaming in areas where that other band class has no coverage. Many industry experts, however, believe that it is preferable for the nationwide public safety broadband network to have more than one commercial partner, to improve network redundancy and capacity, and to promote competition that might reduce costs in the long term.

Many believe that full-spectrum interoperability will, in the long term, maximize the benefits of LTE and LTE Advanced technologies deployed on the 700 MHz band. Coordinating development of 700 MHz band standards among network participants provides an opportunity to maximize the benefits inherent in IP-enabled networks for the safety of the general public. For example, it is possible to create smart phone applications that can link personal mobile devices to emergency command centers, integrating information from those devices into an action plan for response and recovery. Fully implemented within the 700 MHz band, these communications links might help emergency situation managers determine where to most effectively deploy emergency medical service personnel, firefighters, HazMat teams, utility repair crews, or other response and recovery personnel, as appropriate. As a situation stabilizes, evaluations about evacuation routes, shelters, and other post-disaster services could be expedited and information disseminated through wireless and other emergency alert systems.

Whether standards for LTE use within the 700 MHz band can be changed, how they might be changed, and when, are subjects of intense debate within the wireless industry. The debate may be fully resolved only through the transition to LTE Advanced. Development and change in wireless technology are fast-paced. A rapid transition from LTE to LTE Advanced might eliminate separate band classes, negating the need for interoperable work-arounds, some say.

THE FUTURE

One of the goals of effective spectrum management is to create opportunities for the development of innovative technologies. Wireless technology transforms air into desirable services, providing an engine for economic growth and development. The evolution of public safety communications has lagged behind the commercial sector and the military in receiving the benefits of recent innovations. By providing some of the resources needed to build IP-enabled networks for first responders, Congress has created an opportunity to expand the reach and effectiveness of emergency communications for the entire population.

Many experts in advanced communications technology believe that the transition to IP-enabled technologies is likely to bring about the convergence of commercial, military, and emergency response (federal and nonfederal) technologies on common, interoperable platforms.[125] In this view, compatible communications devices will be differentiated by applications developed by stakeholders to meet their mission needs. Infrastructure, spectrum, and mobile devices will be sharable, and it is envisaged that sharing will be encouraged.

The military is linking many of its communications through IP-enabled networks,[126] similar to plans by the public safety community for investment in first responder LTE devices and NG9-1-1. The Department of Defense (DOD) has used the term internetwork to refer to the IP-enabled networks that drive its Global Information Grid (GIG) for network communications.[127] The internetwork, also known as the Convergence Layer, provides analysis and organization of communications activity to facilitate transport. The communications layer that provides the entry and distribution links to services supported by the internetwork is referred to as the Link and Transport Layer by DOD. An Emergency Communications Grid, similar to the military's GIG, might use a common IP-enabled network structure to connect with any type of IP-enabled system, network, or device to support a wide range of services. (See **Figure 2**.)

DOD's internetwork is the equivalent of public safety wireless backhaul, NG9-1-1 network connectivity, or any other network connection that serves the public interest. The Emergency Communications Grid shown in **Figure 2** can also send out emergency alerts to endangered populations,[128] capture information from sensors, manage the Smart Grid to prevent power outages, and support other response and recovery actions. In nonemergency mode, the shared internetwork serves as the conduit for the daily workload of communications. The Emergency Communications Grid represents a unified approach to assuring access and interoperability among all types of communications devices and applications, but it is not envisioned as a single network. The internetwork would be a convergence of many IP-enabled networks that support all the necessary devices and provide the needed links to the Response and Recovery level.

Infrastructure and Spectrum Management

The wireless industry has sometimes compared their licensed radio frequency channels to traffic lanes.[129] These proprietary lanes (access rights assigned for specific frequencies) are used by the vehicles (wireless devices) approved by and often obtained directly from the carrier owning the access rights. Wireless customers have access to their carriers' lanes and usually to other carriers' lanes (roaming privileges) if their vehicles are of the same type used by the lane's owner. Adding more lanes to the highway, that is, providing more spectrum for carriers to expand their networks, is the policy advocated by most representatives of the wireless industry. An increasing number of wireless technology experts are arguing, however, that wireless technology has reached a stage of development where it is possible to allow for a broader variety of vehicles by building a new type of highway. This shift in technology is deemed by many to be a crucial factor in public safety use of commercial technologies. It is argued that as long as there are different classes of wireless devices built to operate on a limited number of predetermined frequencies, the military, public safety, and a number of other users with specific requirements, such as the railroad industry and utilities, will be constrained by high costs distributed among a small number of users.

The successful introduction of the iPhone in 2007 accelerated the convergence of the Internet with mobile technology. The United States is a global leader in wireless technology innovation and adoption.[130] A convergence of technology policy and spectrum policy may be needed to help the country maintain its leadership in innovation. As the nation moves to develop next-generation technologies for public safety communications, wireless industry policies and emergency response and recovery policies may need close coordination to provide a comprehensive safety net forged from sturdy communications links. FirstNet and its private-sector partners have the opportunity to lay the foundation for next-generation communications for public safety.

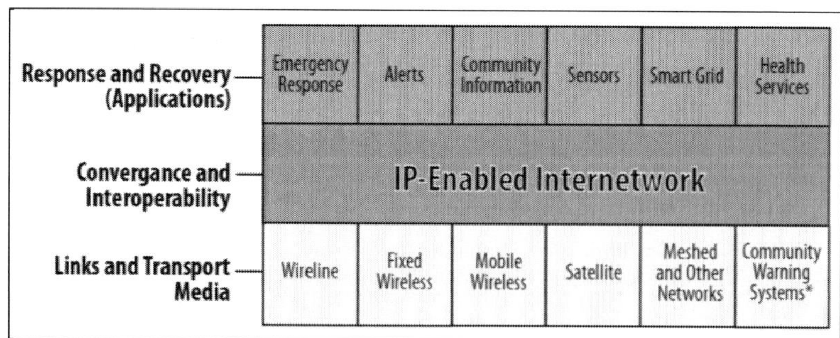

Source: Based on Department of Defense Global Information Grid Architectural Vision; vision for a net-centric, service-oriented, DoD enterprise, prepared by the DOD CIO, June 2007.

* Including sirens, highway and other electronic signs, electronic mailing lists, phone notification systems, and local broadcasting media.

Figure 2. Emergency Communications Grid.

CONSIDERATIONS FOR CONGRESS

FirstNet is required to submit annual reports to Congress containing "a comprehensive and detailed report" of its activities and finances, and to make recommendations for legislative or administrative action that might help FirstNet achieve its goals.[131] Congress may want to pursue additional oversight by, for example, monitoring the participation of states in FirstNet and how state decisions to participate or not participate are affecting network build-out. Congress may also wish to evaluate the level of collaboration in standards development among FirstNet, commercial participants in the network, and federal agencies charged with standards development and certification for network components available to public safety. A comparison of adoption rates for new wireless technologies by the private sector, the military, and public safety agencies may also be of interest to Congress in identifying additional legislative or administrative actions that might further advance the transition to needed new technologies.

End Notes

[1] Hearings in the 112th Congress include Senate, Committee on Science, Commerce and Transportation, "Safeguarding our Future: Building a Nationwide Network for First Responders," February 16, 2011; House, Committee on Homeland Security, "Public Safety Communications: Are the Needs of Our First Responders Being Met?," March 30, 2011; House, Committee on Energy and Commerce, Subcommittee on Communications and Technology, "Using Spectrum to Advance Public Safety, Promote Broadband, Create Jobs, and Reduce the Deficit," April 12, 2011; House, Committee on Energy and Commerce, Subcommittee on Communications and Technology, "Creating an Interoperable Public Safety Network," May 24, 2011; House, Committee on Energy and Commerce, Subcommittee on Communications and Technology, "Legislative Hearing to Address Spectrum and Public Safety Issues," July 15, 2011; House, Committee on Homeland Security, "Hearing on the Attacks of September 11: Where are We Today," September 8, 2011; House, Committee on Energy and Commerce, Subcommittee on Communications and Technology, markup of the Discussion Draft of the Jumpstarting Opportunity with Broadband Spectrum (JOBS) Act of 2011, December 1, 2011.

[2] Discussed in Congressional Research Service General Distribution Memorandum, "Communications Support for Public Safety: The 9/11 Commission Report and Alternative Approaches," by Linda K. Moore, August 25, 2004, and in CRS Report RL31375, *Emergency Communications: Meeting Public Safety Spectrum Needs* by Linda K. Moore, 2002-2003 (out of print; available from the author).

[3] Some of the actions by Congress and by federal agencies were summarized in testimony by Linda K. Moore, Specialist in Telecommunications Policy, Congressional Research Service, before the House Committee on Homeland Security, Subcommittee on Emergency Preparedness, Response, and Communications, "Ensuring Coordination and Cooperation: A Review of Emergency Communications Offices Within the Department of Homeland

Security," November 17, 2011. The GAO has also addressed these issues in reports such as *Emergency Communications: Various Challenges Likely to Slow Implementation of a Public Safety Broadband Network*, February 2012, GAO-12-343 at http://www.gao.gov/assets/590/588795.pdf. CRS reports on the topic include CRS Report R41842, *Funding Emergency Communications: Technology and Policy Considerations*; CRS Report R40859, *Public Safety Communications and Spectrum Resources: Policy Issues for Congress*; CRS Report RL34054, *Public-Private Partnership for a Public Safety Network: Governance and Policy*; CRS Report RL33838, *Emergency Communications: Policy Options at a Crossroads*, all by Linda K. Moore.

[4] P.L. 112-96, Section 6204 (a).

[5] WiFi, for wireless fidelity, operates on unlicensed frequencies that are not assigned to a specific owner but instead are available to support any device approved by the FCC.

[6] Spectrum is segmented into bands of radio frequencies and typically measured in cycles per second, or hertz. Standard abbreviations for measuring frequencies include kHz—kilohertz or thousands of hertz; MHz—megahertz, or millions of hertz; and GHz—gigahertz, or billions of hertz. The 700 MHz band includes radio frequencies from 698 MHz to 806 MHz.

[7] 47 U.S.C. §309 (j) (14).

[8] 763-768 MHz, 793-798 MHz, 768-769 MHz and 798-799 MHz; P.L. 112-96, Section 6201.

[9] 758-763 MHz and 788-793 MHz; P.L. 112-96, Section 6101.

[10] P.L. 112-96, Section 6201.

[11] 769-775 MHz and 799-805 MHz.

[12] P.L. 112-96, Section 6102.

[13] P.L. 112-96, Section 6103.

[14] Metropolitan areas: Boston, MA, Chicago, IL, Dallas/Fort Worth, TX, Houston, TX, Los Angeles, CA, Miami, FL, New York, NY/Newark NJ, Philadelphia, PA, Pittsburgh, PA, San Francisco/Oakland, CA, and Washington, DC.

[15] Details at http://transition.fcc.gov/pshs/public-safety-spectrum/narrowbanding.html.

[16] FCC, "Waiver of Narrowbanding Deadlines for T-Band (470-512 MHz) Licenses," Docket No. WT 99-87, released April 26, 2012.

[17] The FCC has opened a proceeding regarding the 4.9 GHz public safety band, proposing rules and asking for comment on a number of issues to improve spectrum efficiency and encourage greater use of the 4.9 GHz band for public safety broadband communications, http://www.fcc.gov/document/comment-and-reply-comments-date-5th-fnprm-49-ghz-band.

[18] Some cost estimates for building and operating a public safety broadband network are provided in CRS Report R41842, *Funding Emergency Communications: Technology and Policy Considerations*, by Linda K. Moore.

[19] P.L. 112-96, Section 6207.

[20] P.L. 112-96, Section 6413.

[21] P.L. 112-96, Section 6206 (e).

[22] P.L. 112-96, Section 6207 (b).

[23] P.L. 112-96, Section 6206 (b) (4).

[24] P.L. 112-96, Section 6208 (a) (1).

[25] P.L. 112-96, Section 6208 (a) (2).

[26] P.L. 112-96, Section 6208 (a) (3).

[27] P.L. 112-96, Section 6208 (b).

[28] P.L. 112-96, Section 6208 (d).

[29] P.L. 112-96, Section 6212.

[30] P.L. 112-96, Section 6301.

[31] P.L. 112-96, Section 6302 (b).

[32] P.L. 112-96, Section 6302 (a).

[33] *Federal Register*, Vol. 77, No. 162, August 21, 2012, Notice, http://www.ntia.doc.gov/federal-register-notice/2012/development-programmatic-requirements-state-and-local-implementation-gr.

[34] Ibid., Paragraph C, Allowable Grant Activities.
[35] P.L. 112-96, Section 6204.
[36] P.L. 112-96, Section 6206 (a).
[37] Announcement and background information at http://www.ntia.doc.gov/other-publication/2012/acting-secretary-rebecca-blank-announces-board-directors-first-responder-netw.
[38] P.L. 112-96, Section 6205 (a).
[39] P.L. 112-96, Section 6206 (b) (1).
[40] P.L. 112-96, Section 6205 (b) (1).
[41] P.L. 112-96, Section 6206 (b) (4) (D).
[42] P.L. 112-96, Section 6206 (c) (1).
[43] P.L. 112-96, Section 6206 (c) (2).
[44] P.L. 112-96, Section 6206 (c) (3).
[45] P.L. 112-96, Section 6206 (c) (4).
[46] P.L. 112-96, Section 6206 (c) (5).
[47] P.L. 112-96, Section 6206 (c) (7).
[48] See "Statewide Interoperability Coordinators" at http://www.dhs.gov/files/programs/gc_1286986920144.shtm.
[49] P.L. 112-96, Section 6206 (c) (2) (B).
[50] P.L. 112-96, Section 6302 (d).
[51] P.L. 112-96, Section 6302 (e) (1).
[52] P.L. 112-96, Section 6302 (e) (2) and (3).
[53] P.L. 112-96, Section 6302 (e) (3) (C) (iv).
[54] P.L. 112-96, Section 6302 (e) (3) (D).
[55] P.L. 112-96, Section 6302 (f).
[56] P.L. 112-96, Section 6302 (g).
[57] P.L. 112-96, Section 6001 (3).
[58] P.L. 112-96, Section 6204.
[59] P.L. 112-96, Section 6302 (a).
[60] P.L. 112-96, Section 6302 (e) (3) (C) (iii) (I).
[61] P.L. 112-96, Section 6302 (e) (3) (C) (iii) (II).
[62] P.L. 112-96, Section 6201 (b).
[63] P.L. 112-96, Section 6206 (g).
[64] P.L. 112-96, Section 6208 (c).
[65] P.L. 112-96, Section 6209.
[66] P.L. 112-96, Section 6210.
[67] "Smart Grid" is the name given to the evolving electric power network as new information technology systems and capabilities are incorporated. See also CRS Report R41886, *The Smart Grid and Cybersecurity—Regulatory Policy and Issues*, by Richard J. Campbell.
[68] P.L. 112-96, Section 6206 (b) (1) (B).
[69] P.L. 112-96, Section 6206 (b) (1) (C).
[70] P.L. 112-96, Section 6206 (b) (3).
[71] P.L. 112-96, Section 6202 (a).
[72] P.L. 112-96, Section 6202 (b).
[73] P.L. 112-96, Section 6206 (c) (3).
[74] P.L. 112-96, Section 6206 (b) (3).
[75] P.L. 112-96, Section 6203.
[76] P.L. 112-96, Section 6204.
[77] P.L. 112-96, Section 6302 (c).
[78] P.L. 112-96, Section 6205.
[79] P.L. 112-96, Section 6206, (c) (1) and (2).
[80] P.L. 112-96, Section 6206 (c).
[81] P.L. 112-96, Section 6302, (e) (2).
[82] P.L. 112-96, Section 6302, (e) (3) (B).

[83] P.L. 112-96, Section 6302 (e) (3) (C) (i).
[84] P.L. 112-96, Section 6302, (e) (3) (C) (iii).
[85] P.L. 112-96, Section 6302, (e) (3) (D) (i) (III).
[86] "NTIA Official Provides Timelines for Nationwide Broadband Network," by Donny Jackson, urgentcomm.com, March 20, 2012.
[87] P.L. 112-96, Section 6503, "Section 158 "(a).
[88] P.L. 112-96, Section 6503, "Section 158 "(b).
[89] P.L. 112-96, Section 6503, "Section 158 "(c).
[90] P.L. 112-96, Section 6505.
[91] P.L. 112-96, Section 6504.
[92] P.L. 112-96, Section 6509.
[93] P.L. 112-96, Section 6507.
[94] P.L. 112-96, Section 6508.
[95] P.L. 112-96, Section 6506.
[96] P.L. 112-96, Section 6206 (b) (2) (C).
[97] P.L. 112-96, Section 6203 (c) (2).
[98] Also known as 3GPP Release 10, see http://www.3gpp.org/LTE-Advanced.
[99] P.L. 112-96, Section 6206 (b) (1) (A).
[100] P.L. 112-96, Section 6206 (c) (7).
[101] P.L. 112-96, Section 6206 (b) (2) (B).
[102] P.L. 112-96, Section 6203.
[103] P.L. 112-96, Section 6203 (c) (1).
[104] P.L. 112-96, Section 6203 (c) (2).
[105] P.L. 112-96, Section 6203 (c) (1).
[106] P.L. 112-96, Section 6203 (c) (3).
[107] P.L. 112-96, Section 6203 (f).
[108] *Recommended Minimum Technical Requirements to Ensure Nationwide Interoperability for the Nationwide Public Safety Broadband Network*, prepared by the Technical Advisory Board for First Responder Interoperability, Final Report, May 22, 2012, at http://www.fcc.gov/document/recommendations-interoperability-board.
[109] P.L. 112-96, Section 6303 (a).
[110] P.L. 112-96, Section 6303 (b) (1 – 5).
[111] P.L. 112-96, Section 6206 (c) (6).
[112] P.L. 112-96, Section 6206 (b) (2) (A).
[113] House, Committee on Energy and Commerce, Subcommittee on Communications and Technology, "Cybersecurity: Threats to Communications Networks and Public-Sector Responses," testimony of Roberta Stempfley, Acting Assistant Administrator, Office of Cybersecurity and Communications, National Protection and Programs Directorate, Department of Homeland Security, March 28, 2012.
[114] Discussed in GAO report, *Emergency Communications: Various Challenges Likely to Slow Implementation of a Public Safety Broadband Network*, February 2012, GAO-12-343.
[115] In addition to assigning NTIA responsibilities to develop public safety broadband communications, the act also specifies the NTIA's responsibility to promote efficient use of spectrum by the federal government. P.L. 112-96, Section 6410.
[116] More information is available at the PSCR website at http://www.ntia.doc.gov/category/public-safety. PSCR activities were discussed in testimony by Mary H. Saunders, Director, Standards Coordination Office, NIST before the House Committee on Homeland Security, Subcommittees on Emergency Preparedness, Response, and Communications and Cybersecurity, Infrastructure Protection, and Security Technologies, "First Responder Technologies: Ensuring a Prioritized Approach for Homeland Security Research and Development," May 9, 2012.
[117] P.L. 112-96, Section 6401 (b).

[118] The PSCR, for example, has changed its plans for testing public safety interoperability in response to provisions in the act, http://www.pscr.gov/about_pscr/press/broadband/pscr_to_focus_on_public-safety_broadband_interoperability_tests_042012-mission_critical.pdf.

[119] FCC, *Connecting America: The National Broadband Plan*, http://www.broadband.gov/download-plan/.

[120] P.L. 112-96, Section 6211.

[121] P.L. 112-96 Section 6211.

[122] The 3G Partnership Project, known in the United States as 4G Americas, is a consensus-driven, international partnership of industry-based telecommunications standards bodies.

[123] FCC, "Promoting Interoperability in the 700 MHZ Commercial Spectrum," Notice of Proposed Rulemaking, WT Docket No. 12-69, released March 21, 2012.

[124] LTE mobile broadband standard at http://www.3gpp.org/LTE.

[125] See for example reports and meeting discussions of the Visiting Committee on Advanced Technology established by NIST, http://www.nist.gov/director/vcat/.

[126] *Department of Defense Global Information Grid Architectural Vision; vision for a net-centric, service-oriented, DoD enterprise,* prepared by the DOD CIO, June 2007, http://www.dtic.mil/cgi-bin/GetTRDoc?AD=ADA484389.

[127] Ibid., Figure 8, p. 22.

[128] FEMA and the FCC have announced the introduction of the Personal Localized Alerting Network (PLAN) that can deliver geo-targeted text alerts to enabled smart phones. PLAN is the implementation of the Commercial Mobile Alert Service as required by the Warning, Alert, and Response Network Act (WARN Act), P.L. 109-347, §603 (a). FCC Fact Sheet, http://transition.fcc.gov/cgb/consumerfacts/cmas.pdf.

[129] CTIA–The Wireless Association, for example, provides a "Spectrum 101" graphic that uses improvements in transportation as an analogy for changes in the wireless environment, http://files.ctia.org/pdf/ Spectrum_Brochure_111111.pdf.

[130] House Committee on Homeland Security, Subcommittee on Emergency Preparedness, Response, and Communications, "Growing the Wireless Economy Through Innovation," April 18, 2012, spoken and written testimony from various witnesses at the hearing.

[131] P.L. 112-96, Section 6210.

In: Public Safety Broadband ...
Editor: Thomas Giordano

ISBN: 978-1-62417-524-4

Chapter 2

EMERGENCY COMMUNICATIONS: VARIOUS CHALLENGES LIKELY TO SLOW IMPLEMENTATION OF A PUBLIC SAFETY BROADBAND NETWORK*

United States Government Accountability Office

WHY GAO DID THIS STUDY

Emergency responders across the nation rely on land mobile radio (LMR) systems to gather and share information and coordinate their response efforts during emergencies. These public safety communication systems are fragmented across thousands of federal, state, and local jurisdictions and often lack "interoperability," or the ability to communicate across agencies and jurisdictions. To supplement the LMR systems, in 2007, radio frequency spectrum was dedicated for a nationwide public safety broadband network. Presently, 22 jurisdictions around the nation have obtained permission to build public safety broadband networks on the original spectrum assigned for broadband use. This requested report examines (1) the investments in and capabilities of LMR systems; (2) plans for a public safety broadband network and its expected capabilities and limitations; (3) challenges to building this

* This is an edited, reformatted and augmented version of the Highlights of GAO-12-343, a report to congressional requesters, dated February 2012.

network; and (4) factors that affect the prices of handheld LMR devices. GAO conducted a literature review, visited jurisdictions building broadband networks, and interviewed federal, industry, and public safety stakeholders, as well as academics and experts.

What GAO Recommends

The Department of Homeland Security (DHS) should work with partners to identify and communicate opportunities for joint procurement of public safety LMR devices. In commenting on a draft of this report, DHS agreed with the recommendation. GAO also received technical comments, which have been incorporated, as appropriate, in the report.

What GAO Found

After the investment of significant resources—including billions of dollars in federal grants and approximately 100 megahertz of radio frequency spectrum—the current land mobile radio (LMR) systems in use by public safety provide reliable "mission critical" voice capabilities. For public safety, mission critical voice communications must meet a high standard for reliability, redundancy, capacity, and flexibility. Although these LMR systems provide some data services, such as text and images, their ability to transmit data is limited by the channels on which they operate. According to the Department of Homeland Security (DHS), interoperability among LMR systems has improved due to its efforts, but full interoperability of LMR systems remains a distant goal.

Multiple federal entities are involved with planning a public safety broadband network and while such a network would likely enhance interoperability and increase data transfer rates, it would not support mission critical voice capabilities for years to come, perhaps even 10 years or more. A broadband network could enable emergency responders to access video and data applications that improve incident response. Yet because the technology standard for the proposed broadband network does not support mission critical voice capabilities, first responders will continue to rely on their current LMR systems for the foreseeable future. Thus, a broadband network would supplement, rather than replace, current public safety communication systems.

There are several challenges to implementing a public safety broadband network, including ensuring the network's interoperability, reliability, and security; obtaining adequate funds to build and maintain it; and creating a governance structure. For example, to avoid a major shortcoming of the LMR systems, it is essential that a public safety broadband network be interoperable across jurisdictions and devices by following five key elements to interoperable networks: governance, standard operating procedures, technology, training, and usage. With respect to creating a governance structure, pending legislation—the Middle Class Tax Relief and Job Creation Act of 2012, among other things—establishes a new entity, the First Responder Network Authority, with responsibility for ensuring the establishment of a nationwide, interoperable public safety broadband network.

The price of handheld LMR devices is high—often thousands of dollars—in part because market competition is limited and manufacturing costs are high. Further, GAO found that public safety agencies cannot exert buying power in relationship to device manufacturers, which may result in the agencies overpaying for LMR devices. In particular, because public safety agencies contract for LMR devices independently from one another, they are not in a strong position to negotiate lower prices and forego the quantity discounts that accompany larger orders. For similar situations, GAO has recommended joint procurement as a cost saving measure because it allows agencies requiring similar products to combine their purchase power and lower their procurement costs. Given that DHS has experience in emergency communications and relationships with public safety agencies, it is well-suited to facilitate joint procurement of handheld LMR devices.

ABBREVIATIONS

APCO	Association of Public-Safety Communications Officials
BTOP	Broadband Technology Opportunities Program
CAP	Compliance Assessment Program
COPS	Community Oriented Policing Services
Commerce	Department of Commerce
DHS	Department of Homeland Security
DOT	Department of Transportation
GSA	General Services Administration
ECPC	Emergency Communications Preparedness Center

ERIC	Emergency Response Interoperability Center
FEMA	Federal Emergency Management Agency
FCC	Federal Communications Commission
GHz	gigahertz
HHS	Department of Health and Human Services
kHz	kilohertz
LMR	land mobile radio
LTE	Long Term Evolution
MHz	megahertz
NIST	National Institute of Standards and Technology
NTIA	National Telecommunications and Information Administration
NPSTC	National Public Safety Telecommunications Council
OEC	Office of Emergency Communications
OIC	Office of Interoperability and Compatibility
OJP	Office of Justice Programs
P25	Project 25
PSHSB	Public Safety and Homeland Security Bureau
PSCR	Public Safety Communications Research
PSST	Public Safety Spectrum Trust
USDA	Department of Agriculture
UHF	ultra high frequency
VHF	very high frequency

February 22, 2012

Congressional Requesters

Communication systems are essential for public safety officials—especially first responders such as police officers and firefighters—to gather and share information and coordinate their response efforts to save lives during emergencies. Currently, the public safety community uses radio frequency spectrum to transmit and receive critical voice communications through land mobile radio (LMR) systems that are operated by and licensed to state and local jurisdictions.[1] However, such LMR systems can have issues with compatibility, continuity, and capacity in times of large scale emergencies or disasters.[2] In particular, LMR systems often lack "interoperability"—that is, they lack the capabilities that allow first responders to communicate with their counterparts in other agencies and jurisdictions as authorized. Notably, during the terrorist attacks of September 11, 2001, and Hurricane Katrina in 2005, the lack of interoperable public safety communications hampered rescue efforts and the overall effectiveness of public safety operations. Indeed, public safety communication systems are fragmented across thousands of federal, state, and local jurisdictions. This fragmentation hampers operations and puts emergency responders and the public at risk when the responders cannot talk to one another. More than 7 years after the bipartisan 9/11 Commission reported that compatible and adequate communications among public safety organizations at the local, state, and federal levels need to be addressed, the United States still lacks interoperable public safety communications despite substantial investment by emergency response agencies to improve their LMR systems.[3] One factor contributing to continued interoperability issues between jurisdictions is the slow progress in developing standards for the communication devices that operate on these LMR systems.

LMR systems currently have the capacity to handle only minimal data transmissions. To supplement the current LMR systems, plans for a second communication system that would provide nationwide broadband services are underway. Such a network would operate on a different portion of the radio frequency spectrum from the LMR systems. In 2007, the Federal Communications Commission (FCC) assigned a portion of spectrum in the upper 700 megahertz (MHz) band exclusively to public safety for broadband use.[4] Some stakeholders advocated for an additional block of spectrum—known as the "D Block"—to be dedicated to public safety. In March 2008, FCC attempted to auction the D Block with public safety encumbrances but failed to attract a winning commercial bidder.[5] Pending legislation, the Middle Class Tax Relief and Job Creation Act of 2012, includes a provision reallocating the D Block to public safety.[6] Public safety officials believe a public safety broadband network would support important data transmission during emergencies, provide first responders with information not currently available (such as vital signs of critically injured people), and foster greater interoperability. However, FCC has estimated that a stand-alone broadband network would cost approximately $15 billion to construct.[7] Furthermore, due to its technical limitations, the broadband network would not replace the LMR systems for the foreseeable future. A number of jurisdictions sought permission from FCC to begin constructing local or regional public safety broadband networks, and since May 2010, FCC has granted permission to 22 jurisdictions.

Given the important issues surrounding the development of a public safety broadband network, you asked us to provide information about current and planned public safety communication networks and the progress being made to support both voice and data needs. We examined (1) the resources that have been provided for current public safety communication systems and their capabilities and limitations, (2) how a nationwide public safety broadband network is being planned and its anticipated capabilities and limitations, (3) the challenges to building a nationwide public safety broadband network, and (4) the factors that influence competition and cost in the development of public safety communication devices and the options that exist to reduce prices.

To address these objectives, we met with officials and reviewed documentation from 6 of the 22 jurisdictions that received permission from FCC to begin deploying a 700 MHz public safety broadband network, including the San Francisco Bay area, California; Adams County, Colorado; Iowa; Boston, Massachusetts; Mississippi; and Texas. We selected locations based on several criteria, including whether they received federal funding and the size of the planned broadband network. In all of the locations, we interviewed government agencies involved with planning a broadband network, and in locations where planning had progressed, we also interviewed emergency responders who were part of the planning process and vendors selected to build the network, among others. We also interviewed federal agencies involved in public safety communications issues, including entities within the Department of Commerce (Commerce)—the National Institute of Standards and Technology (NIST), the National Telecommunications and Information Administration (NTIA), and the Public Safety Communications Research (PSCR) program; the Departments of Homeland Security (DHS) and Justice; and FCC. In addition, we reviewed relevant documents from these federal entities, including several FCC rulemakings related to developing a broadband network and public safety device competition. We reviewed relevant legislation and conducted a literature review of 43 articles from governmental and academic sources related to emergency communication networks and devices. We interviewed representatives of public safety associations, researchers and consultants recognized for their expertise in public safety communications, and manufacturers of public safety devices, as well as private sector analysts who track this industry. We identified experts and industry stakeholders based on prior published literature, stakeholder recognition and affiliation with the emergency communications and public safety spectrum, and other stakeholders' recommendations. Further details of our scope and methodology are provided in appendix I.

We conducted this performance audit from March 2011 to February 2012 in accordance with generally accepted government auditing standards. Those standards require that we plan and perform the audit to obtain sufficient, appropriate evidence to provide a reasonable basis for our findings and conclusions based on our audit objectives. We believe that the evidence obtained provides a reasonable basis for our findings and conclusions based on our audit objectives.

BACKGROUND

Currently, public safety officials primarily communicate with one another using LMR systems that support voice communication and usually consist of handheld portable radios, mobile radios, base stations, and repeaters, as described:[8]

- Handheld portable radios are typically carried by emergency responders and tend to have a limited transmission range.
- Mobile radios are often located in vehicles and use the vehicle's power supply and a larger antenna, providing a greater transmission range than handheld portable radios.
- Base station radios are located in fixed positions, such as dispatch centers, and tend to have the most powerful transmitters. A network is required to connect base stations to the same communication system.
- Repeaters increase the effective communication range of handheld portable radios, mobile radios, and base station radios by retransmitting received radio signals.

Figure 1 illustrates the basic components of an LMR system.

LMR systems are generally able to meet the unique requirements of public safety agencies. For example, unlike commercial cellular networks, which can allow seconds to go by before a call is set up and answered, LMR systems are developed to provide rapid voice call-setup and group-calling capabilities. When time is of the essence, as is often the case when public safety agencies need to communicate, it is important to have access to systems that achieve fast call-set up times. Furthermore, LMR systems provide public safety agencies "mission critical" voice capabilities—that is, voice capabilities that meet a high standard for reliability, redundancy, capacity, and flexibility. Table 1 describes the key elements for mission critical voice capabilities, as determined by the National Public Safety Telecommunications Council (NPSTC).[9]

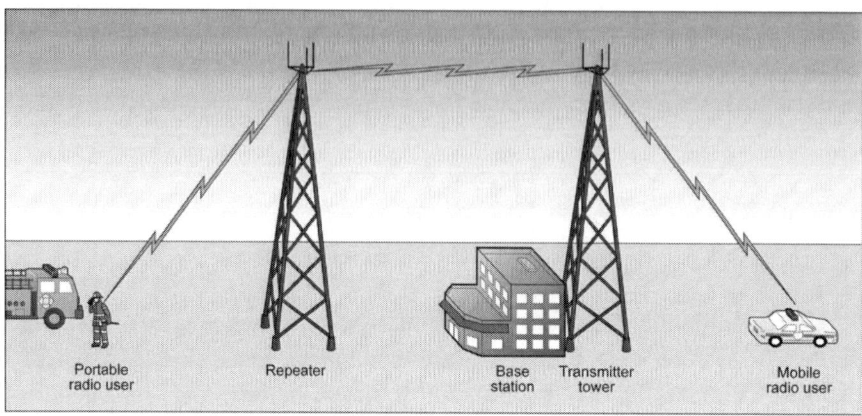

Source: GAO

Figure 1. Depiction of LMR System.

Table 1. Key Elements for Mission Critical Voice Capabilities

Key element	Description
Direct or talk around	Ability to communicate unit-to-unit when out of range of a wireless network or when working in a confined area; both the transmitter and receiver operate without support from infrastructure.
Push-to-talk	Ability to communicate instantly by pushing a button on the device to transmit a voice message. The speaker releases the button to return to a listening mode of operation.
Full duplex voice systems	Ability for multiple users to communicate (talk and listen) at the same time; for example, when communications are necessary with outside parties such as citizens with emergencies, language translation services, and other outside agencies providing service to an incident or event.
Group talk	Ability to communicate on a one-to-many basis. Group talk is of vital importance to the public safety community because it enables a speaker to simultaneously communicate to every member of a group, such as all firefighters in the interior of a burning building.
Talker identification	Ability to identify who is speaking at any given time.
Emergency alerting	Ability to communicate that a life-threatening condition has been encountered and that immediate access to the system is required.
High quality audio	Ability to hear audio in adverse conditions without repetition of the message; for example, an emergency responder must be able to hear prime voice communications regardless of background noises, such as a siren.

Source: GAO based on NPSTC information.

According to NPSTC, for a network to fully support public safety mission critical voice communications, each of the elements in table 1 must address part of the overall voice communications services supported by the network. In other words, NPSTC believes a network cannot be a mission critical network without all of these elements. Furthermore, unlike commercial networks, mission critical communication systems rely on "hardened" infrastructure, meaning that tower sites and equipment have been designed to provide reliable communications even in the midst of natural or man-made disasters. To remain operable during disasters, mission critical communications infrastructure requires redundancy, back-up power, and fortification against environmental stressors such as extremes of temperature and wind.

Nationwide, there are approximately 55,000 public safety agencies. These state and local agencies typically receive a license from FCC to operate and maintain their LMR voice systems. Since these systems are supported by state and local revenues, the agencies generally purchase equipment and devices using their own local budgets without always coordinating their actions with nearby agencies, which can hinder interoperability. Since 1989, public safety associations have collaborated with federal agencies to establish common technical standards for LMR systems and devices called Project 25 (P25). The purpose of these technical standards is to support interoperability between different LMR systems, that is, to enable seamless communication across public safety agencies and jurisdictions. While the P25 suite of standards is intended to promote interoperability by making public safety systems and devices compatible regardless of the manufacturer, it is a voluntary standard and currently incomplete.[10] As a result, many LMR devices manufactured for public safety are not compatible with devices made by rival manufacturers, which can undermine interoperability.

The federal government plays an important role in public safety communications by providing funding for emergency communication systems and working to increase interoperable communication systems. Congress, in particular, has played a critical role by designating radio frequency spectrum for public safety use. Furthermore, Congress can direct action by federal agencies and others in support of public safety. For example, through the Homeland Security Act of 2002, Congress established DHS and required the department, among other things, to develop a comprehensive national incident management system comprising all levels of government and to consolidate existing federal government emergency response plans into a single, coordinated national response plan.[11]

In its regulatory role, FCC licenses all public safety spectrum for state, local, and regional communication networks across the country. This includes more than 134,000 licenses for current public safety LMR narrowband communication systems and the single nationwide public safety broadband license.[12] As part of the digital television transition, Congress, under the Balanced Budget Act of 1997, mandated that FCC allocate 24 MHz of spectrum for public safety use.[13] FCC divided the 24 MHz by assigning 12 MHz for public safety narrowband use and 10 MHz for public safety broadband use.[14] In September 2006, FCC established its Public Safety and Homeland Security Bureau (PSHSB), which is responsible for developing, recommending, and administering FCC's policies pertaining to public safety communications issues. FCC has issued a series of orders and proposed rulemakings and adopted rules addressing how to develop a public safety broadband network, some of which are highlighted:

- In 2007, FCC adopted an order to create a nationwide broadband network with the 10 MHz of spectrum designated for a public safety broadband network and the adjacent 10 MHz of spectrum—the Upper 700 MHz D Block, or "D Block."[15] As envisioned by FCC, this nationwide network would be shared by public safety and a commercial provider and operated by a public/private partnership. However, when FCC presented the D Block for auction in 2008 under these conditions, it received no qualifying bids and thus was not licensed. Subsequently it was found that the lack of commercial interest in the D Block was due in part to uncertainty about how the public/private partnership would work.[16] Although many stakeholders and industry participants called for the D Block to be reallocated to public safety, an alternate view is that auctioning the D Block for commercial use would have generated revenues for the U.S. Treasury.[17] As noted previously, a provision in pending legislation, the Middle Class Tax Relief and Job Creation Act of 2012, reallocates the D Block to public safety.
- In 2007, FCC licensed the 10 MHz of spectrum that FCC assigned for public safety broadband use to the Public Safety Spectrum Trust (PSST), a nonprofit organization representing major national public safety associations. This 10 MHz of spectrum, located in the upper 700 MHz band is adjacent to the spectrum allocated to public safety

for LMR communications. As the licensee, the PSST's original responsibilities included representing emergency responders' needs for a broadband network and negotiating a network sharing agreement with the winner of the D Block auction. However, since the D Block was not successfully auctioned, FCC stayed the majority of the rules guiding the PSST.[18]

- In 2009, public safety entities began requesting waivers from FCC's rules to allow early deployment of broadband networks in the 10 MHz of spectrum licensed to the PSST, and since 2010, FCC granted waivers to 22 jurisdictions for early deployment.[19] These jurisdictions had to request waivers because the rules directing the deployment of a broadband network were not complete. In this report, we refer to the 22 entities receiving waivers as "waiver jurisdictions." As a condition of these waivers, FCC required that local or regional networks would interoperate with each other and that all public safety entities in the geographic area would be invited to use the new networks. In addition, FCC required that all equipment operating on the 700 MHz public safety broadband spectrum comply with Long Term Evolution (LTE), a commercial data standard for wireless technologies.[20] As shown in table 2, of the 22 jurisdictions that successfully petitioned for waivers, only 8 received federal funding. Seven waiver jurisdictions received funding from NTIA's Broadband Technology Opportunities Program (BTOP), a federal grant program authorized through the American Recovery and Reinvestment Act of 2009 that had several purposes, including promoting the expansion of broadband infrastructure.[21]
- In January 2011, FCC adopted rules and proposed further rules to create an effective technical framework for ensuring the deployment and operation of a nationwide, interoperable public safety broadband network.[22] As part of this proceeding, FCC sought comment on technical rules and security for the network as well as testing of equipment to ensure interoperability. The comment period for the proceeding closed on April 11, 2011, and FCC received comments from waiver jurisdictions, consultants, and manufacturers, among others. As of February 7, 2012, FCC did not have an expected issuance date for its final rules.

Table 2. Waiver Jurisdictions as of January 2012 and Federal Funds Provided for Public Safety Broadband Networks

Jurisdiction	Federal funds[a]	Source
Los Angeles Regional Interoperable Communications System (LA-RICS), California	$154.6 million	BTOP
Mississippi Wireless Communications Commission	70.1 million	BTOP
San Francisco, Oakland, and San Jose (BayWEB), California[b]	50.6 million	BTOP
New Jersey	39.6 million	BTOP
New Mexico	38.7 million	BTOP
Charlotte, North Carolina	16.7 million	BTOP
Adams County Communications Center, Colorado	12.1 million	BTOP
Texas	7.6 million	Port Security Grant
	750,000	Regional Catastrophic Planning Grant
Calumet, Outagamie and Winnebago counties, Wisconsin		
Boston, Massachusetts		
Chesapeake, Virginia		
Mesa, Arizona/TOPAZ Regional Wireless Cooperative		
New York, New York		
Pembroke Pines, Florida		
San Antonio, Texas		
Seattle, Washington		
District of Columbia		
Hawaii and various cities and counties		
Iowa		
New York		
Oregon		
Alabama[c]		
Total	**$390.8 million**	

Source: GAO analysis of FCC and NTIA data.

[a] This field appears empty for jurisdictions with no federal funds identified.

[b] Motorola, Inc., is the recipient of the BTOP funds used to build the BayWEB network.

[c] Alabama applied and was granted a waiver but did not execute a lease agreement with the PSST.

In addition to FCC, DHS has been heavily involved since its inception in supporting public safety by assisting federal, state, local, and regional emergency response agencies and policy makers with planning and implementing interoperable communication networks. Within DHS, several divisions have focused on improving public safety communications. DHS also has administered groups that bring together stakeholders from all levels of government to discuss interoperability issues:

- The Emergency Communications Preparedness Center (ECPC) was created in response to Hurricane Katrina by the 21st Century Emergency Communications Act of 2006 to help improve intergovernmental emergency communications information sharing.[23] The ECPC has 14 member agencies with a goal, in part, to support and promote interoperable public safety communications through serving as a focal point and clearing house for information. It has served to facilitate collaboration across federal entities involved with public safety communications.
- SAFECOM is a communications program that provides support, including research and development, to address interoperable communications issues. Led by an executive committee, SAFECOM has members from state and local emergency responders as well as intergovernmental and national public safety communications associations. DHS draws on this expertise to help develop guidance and policy. Among other activities, SAFECOM publishes annual grant guidance that outlines recommended eligible activities and application requirements for federal grant programs providing funding for interoperable public safety communications.

Within Commerce, NTIA and NIST are also involved in public safety communications by providing research support to the PSCR program. The PSCR serves as a laboratory and advisor on public safety standards and technology. It provides research and development to help improve public safety interoperability. For example, the PSCR has ongoing research in many areas related to communications, including the voluntary P25 standard for LMR communication systems, improving public safety interoperability, and the standards and technologies related to a broadband network. PSCR also conducts laboratory research to improve the audio and video quality for public safety radios and devices.

EVEN WITH INVESTMENT OF SIGNIFICANT RESOURCES, CURRENT PUBLIC SAFETY COMMUNICATION SYSTEMS PROVIDE MISSION CRITICAL VOICE CAPABILITIES BUT ARE NOT FULLY INTEROPERABLE

Public Investment

Congress has appropriated billions in federal funding over the last decade to public safety in grants and other assistance for the construction and maintenance of LMR voice communication systems and the purchase of communication devices. Approximately 40 grant programs administered by nine federal agencies have provided this assistance for public safety.[24] Some of the grants provided a one-time infusion of funds, while other grants have provided a more consistent source of funding. For example, in 2007, the one-time Public Safety Interoperable Communications Grant Program awarded more than $960 million to assist state and local public safety agencies in the acquisition, planning, deployment, or training on interoperable communication systems.[25] However, the Homeland Security Grant Program has provided $6.5 billion since 2008, targeting a broad scope of programs that enhance interoperability for states' emergency medical response systems and regional communication systems, as well as planning at the community level to improve emergency preparedness. See appendix II for more information about the grant programs.

State and local governments have also invested millions of dollars of their own funds to support public safety voice communications, and continue to do so. Jurisdictions we visited that received federal grants to support the construction of a broadband network have continued to invest in the upgrade and maintenance of their current LMR voice systems. For example, Adams County, Colorado, has spent about $19.7 million since 2004 on its LMR system, including $6.9 million in local funds, supplemented with $12.8 million in federal grants. Mississippi, another jurisdiction we visited that is constructing a statewide broadband network, has spent about $214 million on its LMR network, including $57 million in general revenue bonds and $157 million in federal grants.[26] Officials in the jurisdictions we contacted stressed the importance of investing in the infrastructure of their LMR networks to maintain the reliability and operability of their voice systems, since it was unclear at what point the broadband networks would support mission critical voice communications. In addition to upgrading and maintaining their LMR networks, many jurisdictions are investing millions of dollars to meet FCC's requirement that communities use their spectrum more efficiently by reducing the bandwidth on which they operate.[27]

In addition to direct federal funding, the federal government has allocated more than 100 MHz of spectrum to public safety over the last 60 years.[28] The spectrum is located in various frequency bands since FCC assigned frequencies to public safety in new bands over time as available frequencies became congested and public safety's need for spectrum increased.[29] Figure 2 displays the spectrum allocated to public safety, which is located between 25 MHz and 4.9 GHz. As noted previously, the Middle Class Tax Relief and Job Creation Act of 2012 requires FCC to reallocate the D Block from commercial use to public safety use.

Public safety agencies purchase radios and communication devices that are designed to operate on their assigned frequency. Since different frequencies of radio waves have different propagation characteristics, jurisdictions typically use the spectrum that is best suited to their particular location. For example, very high frequency (VHF) channels—those located between 30 and 300 MHz—are more useful for communications that must occur over long distances without obstruction from buildings, since the signals cannot penetrate building walls very well. As such, VHF signals are well suited to rural areas. On the other hand, ultra high frequency (UHF) channels—those located between 300 MHz and 3 GHz—are more appropriate for denser urban areas as they have more capacity and can penetrate buildings more easily. When we visited Adams County, Colorado, we learned that public safety officials in the mountainous areas of Colorado use the 150 MHz and 450 MHz bands because of the range of the signals and their ability to navigate around the natural geography. However, public safety officials in the Denver, Colorado, metropolitan area operate on the 700 and 800 MHz frequency bands which can support more simultaneous voice transmissions, such as communications between fire, police, public utility, and transportation officials.

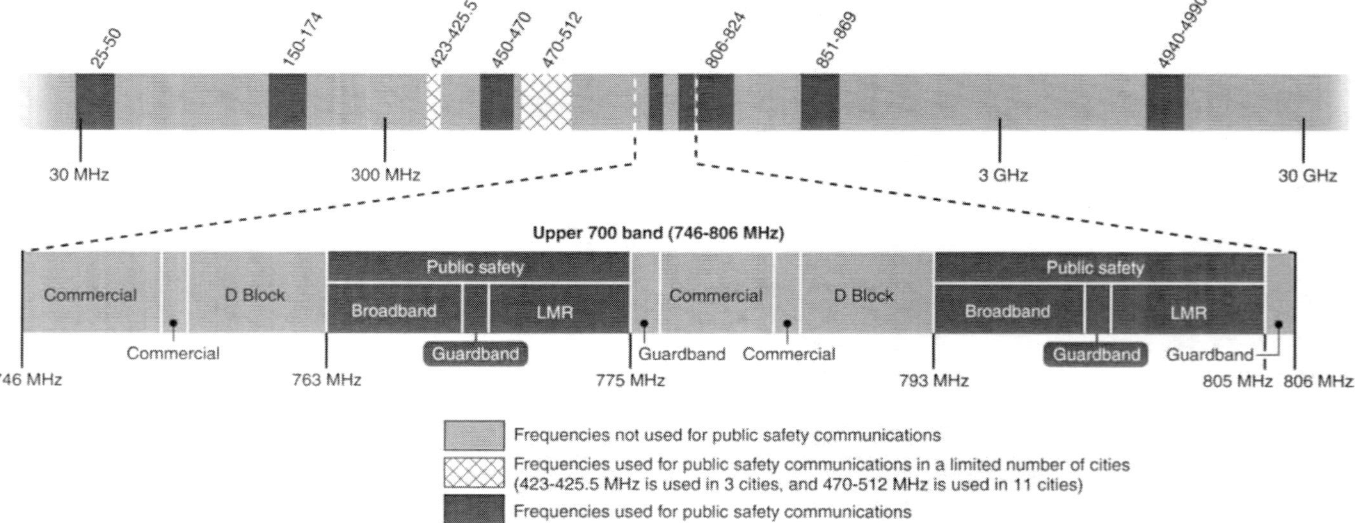

Figure 2. Current Allocation of Public Safety Spectrum.

Current Public Safety Communications Capabilities

The current public safety LMR systems use their allocated spectrum to facilitate reliable mission critical voice communications. Such communications need to be conveyed in an immediate and clear manner regardless of environmental and other operating conditions. For example, while responding to a building fire, firefighters deep within the building need the ability to communicate with each other even if they are out of range of a wireless network. The firefighters are able to communicate on an LMR system because their handheld devices operate on as well as off network. Currently, emergency response personnel rely exclusively on their LMR systems to provide mission critical voice capabilities. One waiver jurisdiction we visited, Mississippi, is constructing a new statewide LMR system and officials there noted a high degree of satisfaction with the planned LMR system. They said the new system is designed to withstand most disasters and when complete, will provide interoperability across 97 percent of the state. Public safety officials in the coastal region of the state have already used the system to successfully respond to problems caused by the Mississippi River flooding in the spring of 2011.

LMR public safety communication systems also are able to provide some data services but the systems are constrained by the narrowband channels on which they operate. These channels allow only restricted data transfer speeds, thus limiting capacity to send and receive data such as text and images, or to access existing databases. Some jurisdictions supplement their LMR systems with commercial data services that give them better access to applications that require higher data transfer rates to work effectively. However, commercial service also has limitations, such as the lack of priority access to the network in an emergency situation.

Interoperability of Current Communication Systems Remains a Limitation

According to DHS, interoperability of current public safety communications has improved as a result of its efforts. In particular, the DHS National Emergency Communications Plan established a strategy for improving emergency communications across all levels of government, and as a result, all states have a statewide interoperability coordinator and governing body to make strategic decisions within the state and guide current and future communications interoperability. According to DHS, it has worked with states to help them evaluate and improve their emergency communications abilities. DHS also helped to develop the Interoperability Continuum, which identifies five critical success elements to assist emergency response agencies and policy makers to plan and implement interoperability solutions for data and voice communications. Furthermore, DHS created guidance to ensure a consistent funding strategy for federal grant programs that allow recipients to purchase communications equipment and enhance their emergency response capabilities. As we have reported in the past, interoperability has also improved due to a variety of local technical solutions.[30] For example, FCC established mutual aid channels, whereby specific channels are set aside for the sole purpose of connecting incompatible systems. Another local solution is when agencies maintain a cache of extra radios that they can distribute during an emergency to other first responders whose radios are not interoperable with their own.

However, despite decades of effort, a significant limitation of current LMR systems is that they are not fully interoperable. One reason for the lack of interoperability is the fragmentation of spectrum assignments for public safety, since existing radios are typically unable to transmit and receive in all these frequencies. Therefore, a rural area using public safety radios operating on VHF spectrum will not be interoperable with radios used in an urban area that operate on UHF spectrum. While radios can be built to operate on multiple frequencies, which could support greater interoperability, this capability can add significant cost to the radios and thus jurisdictions may be reluctant to make such investments. In addition, public safety agencies historically have acquired communication systems without concern for interoperability, often resulting in multiple, technically incompatible radio systems. This is compounded by the lack of mandatory standards for the current LMR systems or devices. Rather, the P25 technical standards remain incomplete and voluntary, creating incompatibility among vendors' products. Furthermore, local jurisdictions are often unable to coordinate to find solutions. Public safety communication systems are tailored to meet the unique needs of individual jurisdictions or public safety entities within a given region. As such, the groups are reluctant to give up management and control of their systems.

PLANNING FOR A NATIONWIDE PUBLIC SAFETY BROADBAND NETWORK PROGRESSES, BUT SUCH A NETWORK WILL NOT SUPPORT MISSION CRITICAL VOICE FOR THE FORESEEABLE FUTURE

Federal Role

Numerous federal entities have helped to plan and begin to define a technical framework for a nationwide public safety broadband network. In particular, FCC, DHS, and Commerce's PSCR program, have coordinated their planning and made significant contributions by developing technical rules, educating emergency responders, and creating a demonstration network, respectively.

Since 2008, FCC has:

- Created a new division within its PSHSB, called the Emergency Response Interoperability Center (ERIC), to develop technical requirements and procedures to help ensure an operable and interoperable nationwide network.[31]
- Convened two advisory committees, the ERIC Technical Advisory Committee and the Public Safety Advisory Committee, that provide advice to FCC.[32] The Technical Advisory Committee's appointees must be federal officials, elected officers of state and local government, or a designee of an elected official. It makes recommendations to FCC and ERIC regarding policies and rules for the technical aspects of interoperability, governance, authentication, and national standards for public safety. ERIC's Public Safety Advisory Committee's members can include representatives of state and local public safety agencies, federal users, and other segments of the public safety community, as well as service providers, equipment vendors, and other industry participants. Its purpose is to make recommendations for a technical framework that will ensure interoperability on a nationwide public safety broadband network.
- Defined technical rules for the broadband network, including identifying LTE as the technical standard for the network, which FCC and public safety agencies believe is imperative to the goal of

achieving an interoperable nationwide broadband network. In addition, FCC sought comments on other technical aspects and challenges to building the network in its most recent proceeding, which FCC hopes will further promote and enable nationwide interoperability. FCC officials said they continue to monitor the waiver jurisdictions that are developing broadband networks to ensure they are meeting the network requirements by reviewing required reports and quarterly filings.

Since 2010, DHS has:

- Partnered with FCC, Commerce, and the Department of Justice to conduct three forums for public safety agencies and others. These forums provided insight about the needs surrounding the establishment of a public safety broadband network as they relate to funding, governance, and the broadband market.
- Coordinated federal efforts on broadband implementation by bringing together the member agencies of ECPC. Also, ECPC updated its grant guidance for federal grant programs to clarify that broadband deployment is an allowable expense for emergency communications grant programs. These updates could result in more federal grant funding going to support the development of a broadband network.
- Updated its SAFECOM program's grant guidance targeting grant applicants to include information pertaining to broadband deployment, based on input from state and local emergency responders.
- Worked with public safety entities to define the LTE standard and write educational materials about the broadband network.
- Partnered with state and regional groups and interoperability coordinators in preparing broadband guidance documentation.
- Represented federal emergency responders and advocated for sharing agreements between the federal government and the PSST that will enable federal users, such as responders from the Federal Emergency Management Agency, to access the broadband network.[33]

Since 2009, PSCR has:

- Worked with public safety agencies to develop requirements for the network and represents their interests before standards-setting organizations to help ensure public safety needs are met.
- Developed a demonstration broadband network that provides a realistic environment for public safety and industry to test and observe public safety LTE requirements on equipment designed for a broadband network. According to PSCR representatives, the demonstration network has successfully brought together more than 40 vendors, including manufacturers and wireless carriers. Among many goals, PSCR aims to demonstrate to public safety how the new technology can meet their needs and encourage vendors to share information and results. FCC requires the 22 waiver jurisdictions and their vendors to participate in PSCR's demonstration network and provide feedback on the challenges they have faced while building the network. PSCR representatives told us that the lessons learned from the waiver jurisdictions would be applied to future deployments.
- Tested interoperable systems and devices and provided feedback to manufacturers. Currently, there are five manufacturers working with PSCR to develop and test systems and devices.

Broadband Network Could Improve Incident Response

With higher data speeds than the current LMR systems, a public safety broadband network could provide emergency responders with new video and data applications that are not currently available. Stakeholders we contacted, including waiver jurisdictions, emergency responders, and federal agencies, identified transmission of video as a key potential capability. For example, existing video from traffic cameras and police car mounted cameras could provide live video feeds for dispatchers. Dispatchers could use the video to help ensure that the proper personnel and necessary equipment are being deployed immediately to the scene of an emergency. Stakeholders we contacted predict that numerous data applications will be developed once a broadband network is complete, and that these applications will have the potential to further enhance incident response. These could range from a global positioning system application that provides directions based on traffic patterns to a 3D graphical floor plan display that supports firefighters' efforts to battle building fires. In addition, unlike the current system, a public safety broadband network could provide access to existing databases of information, such as fire response plans and mug shots of wanted criminals, which could help to keep emergency responders and the public safe. As shown in figure 3, moving from lower bandwidth voice communications to a higher bandwidth broadband network unleashes the potential for the development of a range of public safety data applications.

Besides new applications, a public safety broadband network has the potential to provide nationwide access and interoperability. Nationwide access means emergency responders and other public safety officials could access their home networks from anywhere in the country, which could facilitate a better coordinated emergency response. Interoperability on a broadband network could allow emergency responders to share information irrespective of jurisdiction or type of public safety agency. For example, officials from two waiver jurisdictions indicated that forest fires are a type of emergency that brings together multiple jurisdictions, and in these situations a broadband network could facilitate sharing of response plans. However, an expert we contacted stressed that broadband applications should be tailored to the bandwidth needs of the response task. For example, responders should not use high-definition video when grainy footage would suffice to enable them to pursue a criminal suspect.

Limitations of Broadband Results in Continued Reliance on LMR Voice Systems

A major limitation of a public safety broadband network is that it would not provide mission critical voice communications for many years. LTE, the standard FCC identified for the public safety broadband network, is a wireless broadband standard that is not currently designed to support mission critical voice communications. Commercial wireless providers are currently developing voice over LTE capabilities, but this will not meet public safety's mission critical voice requirements because key elements needed for mission critical voice, such as push-to-talk, are not part of the LTE standard.[34] While one manufacturer believes mission critical voice over LTE will be available as soon as 5 years, some waiver jurisdictions, experts, government officials, and others told us it will likely be 10 years or more due to the challenges described in table 3.

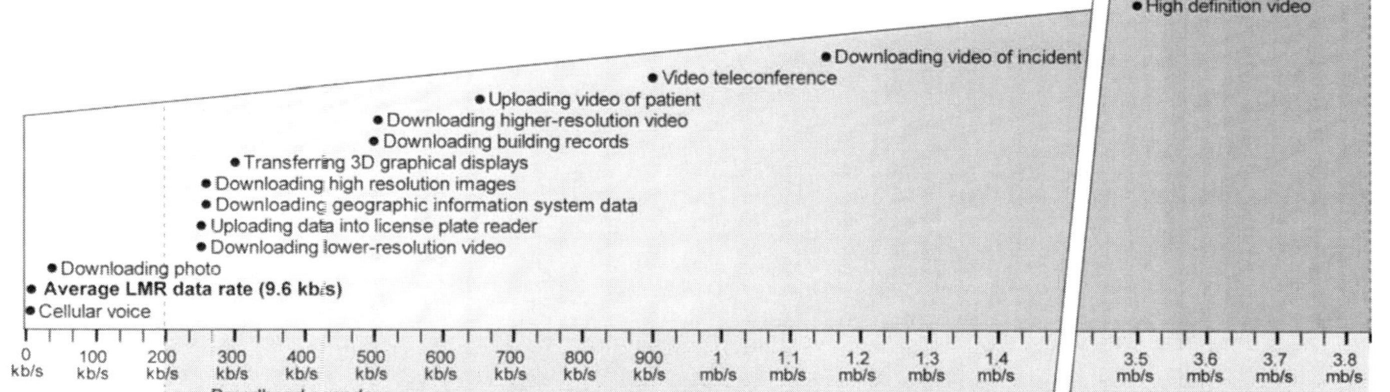

Source: GAO analysis.

Notes: The data transfer rates listed in this figure are estimates. Rates vary depending on whether data are uploaded or downloaded—typically, data are faster to download than to upload. Download speed is the speed of getting information from the Web to a computer or handheld device, and upload speed is the reverse. In addition, higher data transfer rates may be required when data are transmitted to or from a moving device or when a device is further from the tower transmitting the signal.

Kb/s means kilobit per second, and mb/s means megabit per second.

Figure 3. Additional Emergency Response Capabilities from Increased Data Transfer Rates and Bandwidth.

Table 3. Challenges to Developing Mission Critical Voice Capabilities for LTE

Key element for mission critical voice	Challenges for LTE
Direct or talk around	As the LTE standard is a commercial standard, it may not be in the financial interest of commercial providers to develop this capability.
Push-to-talk	Standards need to be developed for this technology and the technology needs to be able to connect and transmit with very little delay.
Group talk	Messages to groups will require bandwidth and the amount can vary based on the size of the group. The bandwidth for voice will take away from the bandwidth available for data.
High quality audio	Digital communications for radio networks have created challenges in the past and the new technology should allow a voice to be understood without repetition, as well as allow background noises to be heard without interference.

Source: GAO based on industry information.

Absent mission critical voice capabilities on a broadband network, emergency responders will continue to rely on their current LMR voice systems, meaning a broadband network would supplement, rather than replace, LMR systems for the foreseeable future. Furthermore, until mission critical voice communications exist, issues that exacerbated emergency response efforts to the terrorist attacks on September 11, 2001—in particular, that emergency responders were not able to communicate between agencies—will not be resolved by a public safety broadband network. As a result, public safety agencies will continue to use devices operating on the current LMR systems for mission critical voice communications, and require spectrum to be allocated for that purpose. Additionally, public safety agencies may be reluctant to give up their LMR devices, especially if they were costly and are still functional. As jurisdictions continue to spend millions of dollars on their LMR networks and devices, they will likely continue to rely on such communication systems until they are no longer functional.[35]

In addition to not having mission critical communications, emergency responders may only have limited access to the public safety broadband network from the interior of large buildings. While the 700 MHz spectrum provides better penetration of buildings than other bands of the spectrum, if emergency responders expect to have access to the network from inside large buildings and underground, additional infrastructure will need to be constructed. For example, antennas or small indoor cellular stations could be installed inside buildings and in underground structures to support access to the network. FCC is seeking comment on this issue as part of its most recent proceeding. Without this added infrastructure, emergency responders using the broadband network may not have access to building blue prints or fire response plans during building emergencies, such as a fire. In fact, one jurisdiction constructing a broadband network that we visited told us their network would not support in-building access in one city of the jurisdiction because the plan did not include antennas for inside the buildings.

A final limitation to a public safety broadband network could be its capacity during emergencies. Emergencies tend to happen in localized areas that may be served by a single cell tower or even a single cellular antenna on a tower.[36] With emergency responders gathering to fight a fire or other emergency, the number of responders and the types of applications in use may exceed the capacity of the network. If the network reaches capacity it could overload and might not send life saving information.[37] Therefore, the network would have to be managed during emergencies to ensure that the most important data are being sent, which could be accomplished by prioritizing data. Furthermore, capacity could be supplemented through deployable cell sites to emergency locations.

Various Challenges Could Jeopardize the Implementation and Functionality of a Public Safety Broadband Network

Although the federal agencies have taken important steps to advance the broadband network, challenges exist that may slow its implementation. Specifically, stakeholders we spoke with prioritized five challenges to successfully building, operating, and maintaining a public safety broadband network. These challenges include (1) ensuring interoperability, (2) creating a governance structure, (3) building a reliable network, (4) designing a secure network, and (5) determining funding sources. FCC, in its *Fourth Further Notice of Proposed Rulemaking*, sought comment on some of these challenges, and as explained further, the challenge of creating a governance structure has been addressed by recent law. However, the other challenges currently remain unresolved and, if left unaddressed, could undermine the development of a public safety broadband network.

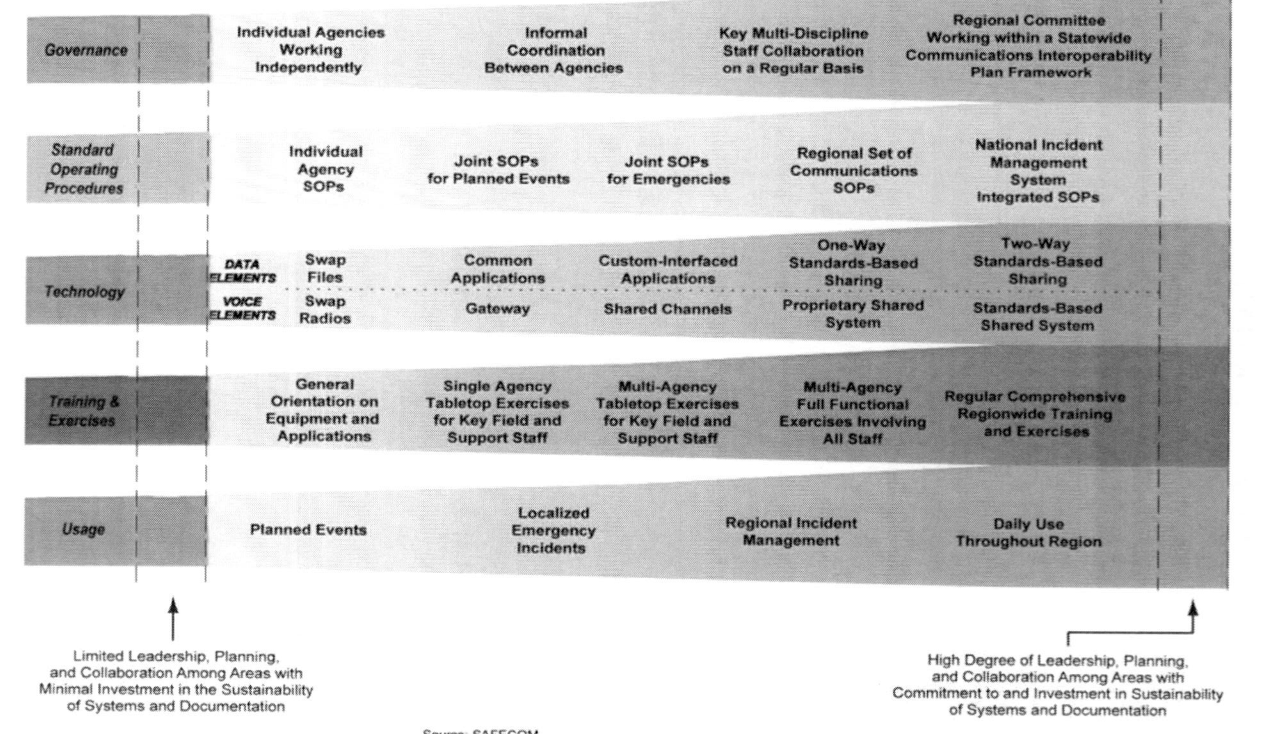

Source: SAFECOM.

Note: This graphic was created by DHS in conjunction with its SAFECOM program and we made a minor revision to its appearance.

Figure 4. SAFECOM Interoperability Continuum with Five Key Elements.

Ensuring interoperability. To avoid a major shortcoming of the LMR communication systems, it is essential that a public safety broadband network be interoperable across jurisdictions and devices. DHS, in conjunction with its SAFECOM program, developed the Interoperability Continuum which identifies five key elements to interoperable networks—governance, standard operating procedures, technology, training, and usage—that waiver jurisdictions and other stakeholders discussed as important to building an interoperable public safety broadband network, as shown in figure 4. For example, technology is critical to interoperability of the broadband network and most stakeholders, including public safety associations, experts, and manufacturers believe that identifying LTE as the technical standard was a good step towards interoperability. To further promote interoperability, stakeholders indicated that additional technical functionality, such as data sharing and roaming capabilities, should be part of the technical design. If properly designed to the technical standard, broadband devices will support interoperability regardless of the manufacturer. Testing devices to ensure they meet the identified standard could help eliminate devices with proprietary applications that might otherwise limit interoperability. In its Fourth Further Notice, FCC solicited input on the technical design of the network and testing of devices to ensure interoperability.[38]

Creating a governance structure. As stated previously, governance is a key element for interoperable networks. A governance authority can promote interoperability by bringing together federal, state, local, and emergency response representatives. Each of the waiver jurisdictions we contacted had identified a governance authority to oversee its broadband network. Jurisdictions we visited, as well as federal agencies, told us that any nationwide network should also have a nationwide governance entity to oversee it. Although several federal entities are involved with the planning of a public safety broadband network, at the time we conducted our work no entity had overall authority to make critical decisions for its development, construction, and operation. According to stakeholders, decisions on developing a common language for the network, establishing user rights for federal agencies, and determining network upgrades, could be managed by such an entity. Pending legislation, the Middle Class Tax Relief and Job Creation Act of 2012, establishes a First Responder Network Authority as an independent authority within NTIA and gave it responsibility for ensuring the establishment of a nationwide, interoperable public safety broadband network. Among other things, the First Responder Network Authority is required to (1) ensure nationwide standards for use and access of the network; (2) issue open, transparent, and competitive requests for proposals to private sector entities to build, operate, and maintain the network; (3) encourage that such requests leverage existing commercial wireless infrastructure to speed deployment of the network; and (4) manage and oversee the implementation and execution of contracts and agreements with nonfederal entities to build, operate, and maintain the network.

Building a reliable network. A public safety broadband network must be as reliable as the current LMR systems but it will require additional infrastructure to do so. As mentioned previously, emergency responders consider the current LMR systems very reliable, in part because they can continue to work in emergency situations. Any new broadband network would need to meet similar standards but, as shown in figure 5, such a network might require up to 10 times the number of towers as the current system. This is because a public safety broadband network is being designed as a cellular network, which would use a series of low powered towers to transmit signals and reduce interference. Also, to meet robust public safety standards, each tower must be "hardened" to ensure that it can withstand disasters, such as hurricanes and earthquakes. According to waiver jurisdictions and other stakeholders, this additional infrastructure and hardening of facilities may be financially prohibitive for many jurisdictions, especially those in rural areas that currently use devices operating on VHF spectrum—spectrum that is especially well suited to rural areas because the signals can travel long distances.

Designing a secure network. Secure communications are important. Designing a protected and trusted broadband network will encourage increased usage and reliance on it. Security for a public safety network will require authentication and access control. By defining LTE as the technical standard for the broadband network, a significant portion of the security architecture is predetermined because the standard governs a certain level of security. Given the importance of this issue, FCC required waiver jurisdictions to include some security features in their networks and FCC's most recent proceeding seeks input on security issues. Furthermore, FCC's Public Safety Advisory Committee has issued a report making several security-related recommendations. For example, it recommended that standardized security features be in place to support roaming to commercial technologies. However, one expert we contacted expressed concern that the waiver jurisdictions were not establishing sufficient network security because they had not received guidance. He believes this would result in waiver jurisdictions using security standards applied to previous networks.

Determining funding sources. It is estimated that a nationwide public safety broadband network could cost up to $15 billion or more to construct, which does not take into account recurring operation and maintenance costs.[39] As noted previously, of the 22 waiver jurisdictions, 8 have received federal grants to support deployment of a broadband network. Some of the other waiver jurisdictions have obtained limited funding from nonfederal sources, such as through issuing bonds. Several of the jurisdictions we spoke with stressed that in addition to the upfront construction costs, the ongoing costs associated with operating, maintaining, and upgrading a public safety broadband network would need to be properly funded. As previously indicated, the ECPC and SAFECOM have updated grant guidance to reflect changing technologies but this does not add additional funding for emergency communications. Rather, it defines broadband as an allowable purpose for emergency communications funding grants that may currently support the existing LMR systems. Since the LMR systems will not be replaced by a public safety broadband network, funding will be necessary to operate, maintain, and upgrade two separate communication systems.

LIMITED COMPETITION AND HIGH MANUFACTURING COSTS INCREASE THE PRICE OF HANDHELD LMR DEVICES, BUT OPTIONS EXIST TO REDUCE PRICES

Competition for Handheld LMR Devices is Limited

Handheld LMR devices often cost thousands of dollars, and many stakeholders, including national public safety associations, state and local public safety officials, and representatives from the telecommunications industry, attribute these high prices to limited competition. Industry analysts and stakeholders estimate that the approximately $4 billion U.S. market for handheld LMR devices consists of one manufacturer with about 75 to 80 percent market share, one or two strong competitors, and several device manufacturers with smaller shares of the market.[40] According to industry stakeholders, competition is weak because of limited entry by device manufacturers; this may be due to (1) the market's relatively small size and (2) barriers to entry that confront nonincumbent device manufacturers.

Small size of the public safety market. The market for handheld LMR devices in the United States includes only about 2 to 3 million customers, or roughly 1 percent of the approximately 300 million customers of commercial telecommunication devices. According to an industry estimate in 2009, approximately 300,000 handheld LMR devices that are P25 compliant are sold each year. Annual sales of handheld LMR devices are small in part because of low turnover. For example, device manufacturers told us that public safety devices are typically replaced every 10 to 15 years, suggesting that less than 10 percent of handheld LMR devices are replaced annually. In contrast, industry and public safety sources indicate that commercial customers replace devices roughly every 2 to 3 years, suggesting that about 33 to 50 percent are replaced annually. Together, low device turnover and a small customer base reduce the potential volume of sales by device manufacturers, which may make the market unattractive to potential entrants.

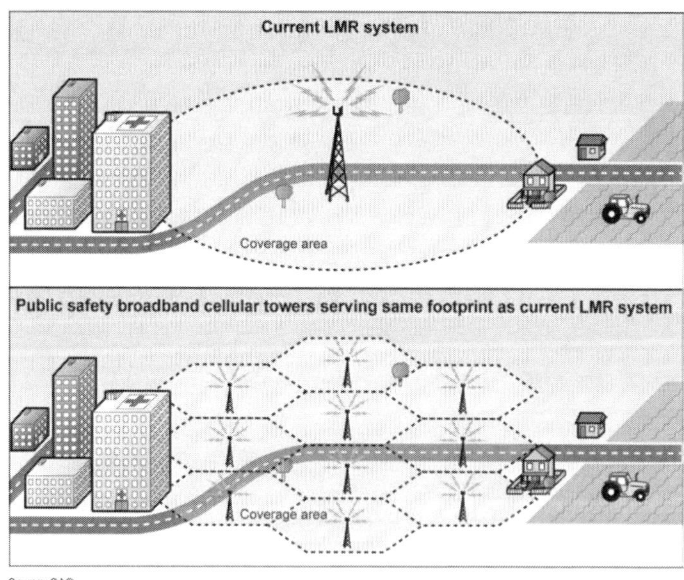

Source: GAO.

Note: The number of towers needed for a cellular network is dependent on the amount of spectrum used by the network and the reliability needed.

Figure 5. Example of Infrastructure Required for Current LMR Systems and a Public Safety Broadband Network.

The size of the market is reduced further by the need for manufacturers to customize handheld LMR devices for individual public safety agencies. Differences in spectrum allocations across jurisdictions have the effect of decreasing the customer base for any single device. As previously discussed, public safety agencies operate on different frequencies scattered across the radio spectrum. For example, one jurisdiction may need devices that operate on 700 MHz frequencies, whereas another jurisdiction may need devices that operate on both 800 MHz and 450 MHz frequencies. Existing handheld LMR devices typically do not transmit and receive signals in all public safety frequencies. As a result, device manufacturers cannot sell a single product to customers nationwide, and must tailor devices to the combinations of frequencies in use by the purchasing agency.

Barriers to entry by nonincumbent manufacturers. Device manufacturers wishing to enter the handheld LMR device market face barriers in doing so, which further limits competition. The use of proprietary technologies represents one barrier to entry. The inclusion of proprietary technologies often makes LMR devices noninteroperable with one another. This lack of interoperability makes it costly for customers to switch the brand of their devices, since doing so requires them to replace or modify older devices. These switching costs may continually compel customers to buy devices from the incumbent device manufacturer, preventing less established manufacturers from making inroads into the market. For example, in a comment filed with FCC, one of the jurisdictions we visited said that device manufacturers offer a proprietary encryption feature for free or at only a nominal cost.[41] When a public safety agency buys devices that incorporate this proprietary encryption feature, the agency cannot switch its procurement to a different manufacturer without undertaking costly modifications to its existing fleet of devices. Switching costs are particularly high when a device manufacturer has installed a communication system that is incompatible with competitors' devices. In this scenario, a public safety agency cannot switch to a competitor's handheld device without incurring the cost of new equipment or a patching mechanism to resolve the incompatibility. Even where devices from different manufacturers are compatible, a fear of incompatibility may deter agencies from switching to a nonincumbent brand. According to industry stakeholders—and as we have confirmed in the past—devices marketed as P25 compliant often are not interoperable in practice.[42] This lack of confidence in the P25 standard may encourage agencies to continue buying handheld LMR devices from their current brand, placing less established device manufacturers at a disadvantage and thus discouraging competition.

At the same time that less established manufacturers are at a disadvantage, the market leader enjoys distinct "incumbency advantages." These advantages refer to the edge that a manufacturer derives from its position as incumbent, over and above whatever edge it derives from the strength of its product:

- According to an industry analyst, some public safety agencies are reluctant to switch brands of handheld LMR devices because their emergency responders are accustomed to the placement of the buttons on their existing devices.
- According to another industry analyst, the extensive network of customer representatives that the market leader has established over time presents an advantage. According to this analyst, less established device manufacturers face difficulty winning contracts because their networks of representatives are comparatively thin.
- The well-recognized brand of the market leader also represents an advantage. According to one stakeholder, some agencies mistakenly believe that only the market leader is able to manufacturer devices compliant with P25, and thus conduct sole-source procurements with this manufacturer. Even where procurements are competitive, the market leader is likely to enjoy an upper hand over its competitors; according to an industry analyst, local procurement officers prefer to buy handheld LMR devices from the dominant device manufacturer because doing so is an uncontroversial choice in the eyes of their management.[43]

High Manufacturing Costs and Lack of Buying Power Increase Device Prices

Competition aside, handheld LMR devices are costly to manufacture, so their prices will likely exceed prices for commercial devices regardless of how much competition exists in the market. First, this is in large part because these devices need to be reinforced for high-pressure environments. Handheld LMR devices must be able to withstand extremes of temperature as well physical stressors such as dust, smoke, impact, and immersion in water. Second, they also have much more robust performance requirements than commercial devices—including greater transmitter and battery power—to enable communication at greater ranges and during extended periods of operation. Third, the devices are produced in quantities too small to realize the cost savings of mass production. Manufacturers of commercial telecommunication devices can keep prices lower simply because of the large quantities they produce. For example, one industry stakeholder told us that economies of scale begin for commercial devices when a million or more devices are produced per manufacturing run. In contrast, LMR devices are commonly produced in manufacturing runs of 25,000 units. Fourth, the exterior of handheld LMR devices must be customized to the needs of emergency responders. For example, the buttons on these devices must be large enough to press while wearing bulky gloves.

In addition, given that the P25 standard remains incomplete and voluntary, device manufacturers develop products based on conflicting interpretations of the standard, resulting in incompatibilities between their products. Stakeholders from one jurisdiction we visited said that agencies can request add-on features—such as the ability to arrange channels according to user preference or to scan for radio channels assigned for particular purposes—which fall outside the P25 standard. These features increase the degree of customization required to produce handheld LMR devices, pushing costs upward.

Furthermore, public safety agencies may be unable to negotiate lower prices for handheld LMR devices because they cannot exert buying power in relationship with device manufacturers. We found that public safety agencies are not in an advantageous position to negotiate lower prices because they often request customized features and negotiate with device manufacturers in isolation from one another. According to a public safety official in one jurisdiction we contacted, each agency has unique ordinances, purchasing mechanisms, and bidding processes for devices. Because public safety agencies contract for handheld LMR devices in this independent manner, they sacrifice the quantity discounts that come from placing larger orders. Moreover, they are unlikely to know what other agencies pay for similar devices, enabling device manufacturers to offer different prices to different jurisdictions rather than set a single price for the entire market. One public safety official told us that small jurisdictions therefore pay more than larger jurisdictions for similar devices. As we have reported in the past, agencies that require similar products can combine their market power—and therefore obtain lower prices—by engaging in joint procurement.[44] Therefore, wider efforts to coordinate procurement at the state, regional, or national level are likely to increase the buying power of public safety agencies and help bring down prices.

Although these factors drive up prices in the current market for handheld LMR devices, industry observers said that many of these factors diminish in the future market for handheld broadband devices. As described earlier, FCC has mandated a commercial standard, LTE, for devices operating on the new broadband networks. The use of this standard may reduce the prevalence of proprietary features that inhibit interoperability. In addition, the new broadband networks will operate on common 700 MHz spectrum across the nation, eliminating the need to customize devices to the frequencies in use by individual jurisdictions. Together, the adoption of a commercial standard and the use of common spectrum are likely to increase the uniformity of handheld public safety devices, which in turn is likely to strengthen competition and enable the cost savings that come from bulk production. In addition, industry analysts and federal officials told us that they expect a heightened level of competition in the market for LTE devices because multiple device manufacturers are expected to develop them.

Options Exist to Reduce the Prices of Handheld LMR Devices

Options exist to reduce prices in the market for handheld LMR devices by increasing competition and the bargaining power of public safety agencies. One option is to reduce barriers to entry into the market. As described above, less established manufacturers may be discouraged from entering the market for handheld LMR devices because of the lack of interoperability between devices produced by different manufacturers. Consistent implementation of the P25 standard would increase interoperability between devices, enabling public safety agencies to mix and match handheld LMR devices from different brands. As we have reported in the past, independent testing is necessary to ensure compliance with standards and interoperability among products.[45] In the past several years, NIST and DHS have established a Compliance

Assessment Program (CAP) for the P25 standard. CAP provides a government-led forum in which to test devices for conformance with P25 specifications.[46] If the CAP program succeeds in increasing interoperability, it may reduce switching costs—that is, the expense of changing manufacturers—and thus may open the door to greater competition. Although CAP is a promising means to lower costs in this way, it is too soon to assess its effectiveness.

A second option is for public safety agencies to engage in joint procurement to lower costs. Joint procurement of handheld LMR devices could increase the bargaining power of agencies as well as facilitate cost savings through quantity discounts. One public safety official we interviewed said that while local agencies seek to maintain control over operational matters—such as which emergency responders operate on which channels—they are likely to cede control in procurement matters if doing so lowers costs.[47] As described earlier in this report, DHS provides significant grant funding, technical assistance, and guidance to enhance the interoperability of LMR systems. For example, as described in its January 2012 Technical Assistance Catalog, DHS's Office of Emergency Communications supports local public safety entities to ensure that LMR design documents meet P25 specifications and are written in a vendor-neutral manner. Based on its experience in emergency communications and its outreach to local public safety representatives, DHS is positioned to facilitate and incentivize opportunities for joint procurement of handheld LMR devices.

CONCLUSIONS

Despite their interoperability limitations, traditional LMR systems have provided public safety agencies with mission critical voice capabilities that commercial broadband systems cannot provide. These LMR systems will continue to be essential for public safety communications until broadband systems are able to meet public safety requirements, particularly for mission critical voice. As a result, a public safety broadband network would likely supplement, rather than replace, current LMR systems for the foreseeable future. Although a public safety broadband network could enhance incident response, it would have limitations and be costly to construct. Furthermore, since the LMR systems will still be operational for many years, funding will be necessary to operate, maintain, and upgrade two separate public safety communication systems.

At the time of our work, there was not an administrative entity that had the authority to plan, oversee, or direct the public safety broadband spectrum. As a result, overarching management decisions had not been made to guide the development or deployment of a public safety broadband network. According to SAFECOM's interoperability continuum, governance structures provide a framework for collaboration and decision making with the goal of achieving a common objective and therefore foster greater interoperability. In addition to ensuring interoperability, a governance entity with proper authority could help to address the challenges identified in this report, such as ensuring the network is secure and reliable. Pending legislation, the Middle Class Tax Relief and Job Creation Act of 2012, establishes an independent authority within NTIA to manage and oversee the implementation of a nationwide, interoperable public safety broadband network.

Handheld communication devices used by public safety officials can cost thousands of dollars, mostly due to limited competition and high manufacturing costs. However, public safety agencies also lack buying power vis-à-vis the device manufacturers, which may result in the agencies overpaying for the devices. In particular, since public safety agencies negotiate individually with device manufacturers, they are unlikely to know what other agencies pay for comparable devices and they sacrifice the increased bargaining power and economies of scale that accompany joint purchasing. Especially in rural areas, public safety agencies may be overpaying for handheld devices. We have repeatedly recommended joint procurement as a cost saving measure for situations where agencies require similar products because it allows them to combine their market power and lower their procurement costs. Given that DHS has expertise in emergency communications and relationships with local public safety representatives, we believe it is well-suited to facilitate opportunities for joint procurement of handheld communication devices.

RECOMMENDATION FOR EXECUTIVE ACTION

To help ensure that public safety agencies are not overpaying for handheld communication devices, the Secretary of Homeland Security should work with federal and state partners to identify and communicate opportunities for joint procurement of public safety LMR devices.

AGENCY COMMENTS AND OUR EVALUATION

We provided a draft of this report to Commerce, DHS, the Department of Justice, and FCC for their review and comment. In the draft report we sent to the agencies, we included a matter for congressional consideration for ensuring that a public safety broadband network has adequate direction and oversight, such as by creating a governance structure that gives authority to an entity to define rules and develop a plan for the overarching management of the network. As a result of pending legislation that addresses this issue, we removed the matter for congressional consideration from the final report.

Commerce provided written comments in which it noted that NIST and NTIA will continue to collaborate with and support state, local, and tribal public safety agencies and other federal agencies to help achieve effective and efficient public safety communications.

In commenting on the draft report, DHS concurred with our recommendation that it should work with federal and state partners to identify and communicate opportunities for joint procurement of public safety LMR devices. While DHS noted that this recommendation will not likely assist near-term efforts to implement a public safety broadband network, assisting efforts for the broadband network was not the intention of the recommendation. Rather, we intended this recommendation to help ensure that public safety agencies do not overpay for handheld LMR devices by encouraging joint procurement. DHS suggested in response to our recommendation that a GSA solution may be more appropriate than DHS contracting activity. Although we recognize that a GSA solution is one possibility for joint procurement of handheld LMR devices, other opportunities and solutions might exist. We believe DHS, based on its experience in emergency communications and its outreach to state and local public safety representatives, is best suited to identify such opportunities and solutions for joint procurement and communicate those to the public safety agencies. In its letter, DHS also noted that it continues to work with federal, state, local, and private-sector partners to facilitate the deployment of a nationwide public safety broadband network, and stressed that establishing an effective governance structure is crucial to ensuring interoperability and effective use of the network.

Commerce, DHS, the Department of Justice, and FCC provided technical comments on the draft report, which we incorporated as appropriate.

Mark L. Goldstein
Director, Physical Infrastructure

APPENDIX I. OBJECTIVES, SCOPE, AND METHODOLOGY

This report examines current communication systems used by public safety and issues surrounding the development of a nationwide public safety broadband network. Specifically, we reviewed (1) the resources that have been provided for current public safety communication systems and their capabilities and limitations, (2) how a nationwide public safety broadband network is being planned and its anticipated capabilities and limitations, (3) the challenges to building a nationwide public safety broadband network, and (4) the factors that influence competition and cost in the development of public safety communication devices and the options that exist to reduce prices.

To address all objectives, we conducted a literature review of 43 articles from governmental and academic sources on public safety communications. We reviewed these articles and recorded relevant evidence in workpapers, which informed our report findings. To identify existing studies, we conducted searches of various databases, such as EconLit, ProQuest, Academic OneFile, and Social SciSearch. We also pursued a snowball technique—following citations from relevant articles—to find other relevant articles and asked external researchers that we interviewed to recommend additional studies. These research methods produced 106 articles for initial review. We vetted this initial list by examining summary level information about each piece of literature, giving preference to articles that appeared in peer-reviewed journals and were germane to our research objectives. As a result, the 43 studies that we selected for our review met our criteria for relevance and quality. For the 13 articles related to our fourth objective—factors that affect competition and cost in the market for public safety communication devices—a GAO economist performed a secondary review and confirmed the relevance to our objective. Articles were then reviewed and evidence captured in workpapers. The workpapers were then reviewed for accuracy of the evidence gathered. We performed these searches and identified articles from June 2011 to September 2011.

We also interviewed government officials or stakeholders in 6 of the 22 jurisdictions that are authorized to build early public safety broadband networks and obtained information concerning each objective. In particular, we obtained information concerning their current communication systems and its capabilities, including any funding received to support the current network. We discussed their plan for building a public safety broadband network and the challenge they had faced thus far, including the role each thought the federal government should play in developing a network. We also discussed their views on the communication device market and the factors shaping the market. We selected jurisdictions to contact based on three criteria: (1) whether the jurisdiction received grant funds from the National Telecommunications and Information Administration (NTIA) to help build the network, (2) whether the planned network would be a statewide or regional network, and (3) geographic distribution across the nation. Table 4 lists the jurisdictions we selected based on these criteria. We selected jurisdictions based on NTIA grant funding because these jurisdictions had received the most significant federal funds dedicated towards developing a broadband network. Other jurisdictions either had not identified any funding or applied smaller grant funding that was not primarily targeted at emergency communications. We selected the size of the network, statewide or regional, to determine if challenges differed based on the size of the network and the number of entities involved. Finally, we selected sites based on the geographic region to get a geographic mix of jurisdictions from around the country. In jurisdictions that received NTIA funding, we met with government officials and emergency responders.[48] In jurisdictions that did not receive NTIA funding we met with the government officials since the network had not progressed as much.

Table 4. Jurisdictions Contacted

Jurisdiction	Criteria		
	NTIA funding	**Scope of network**	**Geographic region**
Boston, Massachusetts	No	Regional	East
Texas	No	Statewide	South
Iowa	No	Statewide	Midwest
San Francisco, Oakland, and San Jose (BayWEB), California	Yes	Regional	West
Adams County Communications Center, Colorado	Yes	Regional	West
Mississippi Wireless Communications Commission	Yes	Statewide	South

Source: GAO analysis.

To determine the resources that have been provided for current public safety communication systems, we reviewed Federal Communications Commission (FCC) data on spectrum allocations for land mobile radio (LMR) systems. In addition, we reviewed relevant documentation and interviewed officials from offices within the Departments of Commerce (Commerce), Homeland Security (DHS), and Justice that administer grant programs or provide grants that identify public safety communications as an allowable expense. We selected these agencies to speak with because they had more grant programs providing funds or were regularly mentioned in interviews as providing funds for public safety communications. We also reviewed documents from agencies, such as the Departments of Agriculture and Transportation, which similarly operate grant programs that identify public safety communications as an allowable expense. The grants were identified by DHS's SAFECOM program as grants that can support public safety communications.

To identify the capabilities and limitations of current public safety communication systems, we reviewed relevant congressional testimonies, academic articles on the capabilities and limitations of LMR networks, and relevant federal agency documents, including DHS's National Emergency Communications Plan. We interviewed officials from three national public safety associations—the Association of Public-Safety Communications Officials (APCO), National Public Safety Telecommunications Council (NPSTC), and the Public Safety Spectrum Trust (PSST)—as well as researchers and consultants referred to us for their knowledge of public safety communications and identified during the literature review process.

To determine the plans for a nationwide public safety broadband network and its expected capabilities and limitations, we reviewed relevant congressional testimonies and academic articles on services and applications likely to operate on a public safety broadband network, the challenges to building, operating, and maintaining a network. We interviewed officials from APCO, NPSTC, and PSST, as well as researchers and consultants who specialize in public safety communications to understand the potential capabilities of the network. In addition, we reviewed FCC orders and notices of proposed rulemaking relating to broadband for public safety, as well as comments on this topic submitted to FCC.

To determine the federal role in the public safety broadband network, we interviewed multiple agencies involved in planning this network. Within FCC, we interviewed officials from the Public Safety and Homeland Security

Bureau (PSHSB), the mission of which is to ensure public safety and homeland security by advancing state-of-the-art communications that are accessible, reliable, resilient, and secure, in coordination with public and private partners. Within Commerce, we interviewed officials from NTIA and the National Institute of Standards and Technology (NIST), two agencies that develop, test, and advise on broadband standards for public safety. We also interviewed officials from the Public Safety Communications Research (PSCR) program, a joint effort between NIST and NTIA that works to research, develop, and test public safety communication technologies. Within DHS, we interviewed officials from the Office of Emergency Communications (OEC) and the Office of Interoperability and Compatibility (OIC), two agencies that provide input on the public safety broadband network through their participation on interagency coordinating bodies.

To determine the technological, historical, and other factors that affect competition in the market for public safety devices, as well as what options exist to reduce the cost of these devices, we reviewed the responses to FCC's notice seeking comment on competition in public safety communications technologies. In addition, we reviewed our prior reports and correspondence on this topic between FCC and the House of Representatives Committee on Energy and Commerce that occurred in June and July of 2010 and April and May of 2011. We also conducted an economic literature review that included 13 academic articles examining markets for communications technology and, in particular, how issues of standards, compatibility, bundling, and price discrimination affect entry and competition in these markets. These articles provided a historical and theoretical context for communication technology markets, which helped shape our findings. We asked about factors affecting the price of public safety devices, as well as how to reduce these prices, during our interviews with national public safety organizations, local and regional public safety jurisdictions, and the federal agencies we contacted during our audit work. We also interviewed two researchers specifically identified for their knowledge of communication equipment markets based on their congressional testimony or publication history. In addition, we interviewed representatives from four companies that produce public safety devices or network components, as well as two financial analysts who track the industry.

We conducted this performance audit from March 2011 to February 2012 in accordance with generally accepted government auditing standards. Those standards require that we plan and perform the audit to obtain sufficient, appropriate evidence to provide a reasonable basis for our findings and conclusions based on our audit objectives. We believe that the evidence obtained provides a reasonable basis for our findings and conclusions based on our audit objectives.

APPENDIX II. FEDERAL GRANT PROGRAMS FOR EMERGENCY COMMUNICATIONS

SAFECOM, a program administered by DHS, has identified federal grant programs across nine agencies, including the Departments of Agriculture, Commerce, Education, Health and Human Services, Homeland Security, Interior, Justice, Transportation, and the U.S. Navy that allow grant funds to fund public safety emergency communications efforts. These grants include recurring grants that support emergency communications, research grants that fund innovative and pilot projects, and past grants that may be funding ongoing projects. While the funding from these grants can support emergency communications, the total funding reported does not mean it was all spent on emergency communications.[49] We provided the amounts of the grants and the years funded when this information was available.

Department of Commerce

Two agencies within Commerce—NTIA and NIST—administer grants that allow funds to be directed towards public safety emergency communications (see table 5).

Table 5. Commerce Grants with Emergency Communications as an Allowable Expenditure

Program (administering agency)	Description	Years funded (if available)
Broadband Technology Opportunities Program (NTIA)	The American Recovery and Reinvestment Act of 2009 provided one-time funding for improvements to broadband access, as well as, broadband education, awareness, training, equipment, and support to community anchor institutions, among other purposes. This program provided $3.9 billion including more than $382 million for infrastructure projects to deploy public safety wireless broadband networks.	Fiscal years 2009-2010

Program (administering agency)	Description	Years funded (if available)
Measurement Science and Engineering Research Grants: Electronics and Electrical Engineering Laboratory Program (NIST)	The program provides grants and cooperative agreements for the development of fundamental electrical metrology and of metrology supporting industry and government agencies (including law enforcement standards).	n/a
Measurement Science and Engineering Research Grants: Information Technology Laboratory Grants Program (NIST)	The program provides grants and cooperative agreements in the broad areas of mathematical and computational sciences, advanced network technologies, information access, and software testing.	n/a
Measurement Science and Engineering Research Grants: Physical Measurement Laboratory Grants Program[a] (NIST)	This program provides grants and cooperative agreements in several fields of research, including research on time and frequency standards and applications.	n/a
Public Safety Interoperable Communications Grant Program (NTIA and the Federal Emergency Management Agency)	This grant program provided one-time funding to states and territories to enable and enhance public safety agencies' interoperable communications capabilities.[b] The program awarded more than $968 million to fund interoperable communications projects in the 56 states and territories.	Fiscal year 2007

Source: DHS and Commerce.

Note: Not available is referenced as n/a in the table.

[a] Formerly the Measurement Science and Engineering Research Grants: Physics Laboratory Grant Program.

[b] The Public Safety Interoperable Communications (PSIC) Grant Program was created by the Deficit Reduction Act of 2005 (Pub. L. No. 109-171, §3006, 120 Stat. 4, 24 (2006)). The Digital Television Transition and Public Safety Act of 2005 allowed funding for the Interoperable Emergency Communications Grant Program. (Pub. L. No. 111-96).

Department of Homeland Security

Two agencies within DHS administer grants that allow funds to be directed towards public safety emergency communications—the Federal Emergency Management Agency (FEMA) and the Science and Technology Directorate. Another agency, OEC, has administered one such grant program. Furthermore, DHS maintains an authorized equipment list to document equipment eligible for purchase under its grant programs, including interoperable communications equipment.[50] (See table 6.)

Table 6. DHS Grants with Emergency Communications as an Allowable Expenditure

Program	Description	Years funded (if available)
Assistance to Firefighters Grant Program (FEMA)	This program awards grants to fire departments to enhance their ability to protect the public and fire service personnel from fire and fire-related hazards.	n/a
Border Interoperability Demonstration Project (OEC)	This was a one-time competitive demonstration project that provided funding to state, local, and tribal jurisdictions to develop and identify innovative approaches to improving interoperable emergency communications along and across U.S. international borders. This grant provided $25.6 million to eligible jurisdictions.	Fiscal year 2010
Buffer Zone Protection Program (FEMA)	This program provides funding to states for improving preparedness capabilities of jurisdictions surrounding high-priority critical infrastructure, such as nuclear power plants and financial institutions. This program has provided almost $333 million to eligible jurisdictions.	Fiscal years 2005-2010
Citizen Corps Grant Program (FEMA)	This program's mission is to bring community and government leaders together to coordinate community-based planning efforts. This program has provided almost $186 million.	Fiscal years 2007-2011
Emergency Management Performance Grant (FEMA)	The grant assists state and local governments in enhancing and sustaining their all-hazards emergency management capabilities.	n/a
Emergency Management Performance Grants Supplemental (FEMA)	This supplemental grant provided an additional $50 million for the Emergency Management Performance Grant program.	Fiscal year 2007
Emergency Operations Center Grant Program (FEMA)	This grant program is intended to help building or renovating state, local, or tribal Emergency Operation Centers.	n/a
Intercity Bus Security Grant Program (FEMA)	The purpose of this program is to provide funding to protect the intercity bus systems and people traveling on the systems from terrorism. Operators may use the funds to	n/a

Program	Description	Years funded (if available)
	purchase emergency communications technology that focuses on theft prevention, real-time bus inventory, tracking, monitoring, and locating technologies.	
Intercity Passenger Rail–Amtrak (FEMA)	The purpose this grant program is to protect critical passenger rail infrastructure and the traveling public from terrorism. Amtrak is the only entity eligible to apply for funding under this grant program.	n/a
Interoperable Emergency Communications Grant Program (FEMA)	This program provided funding to state, local and tribal entities for governance, planning, and training to improve interoperable emergency communications.[a]	n/a
Long-Range Broad Agency Announcement (Science and Technology)	This program serves as an open invitation to the scientific and technical communities to fund pioneering research and development projects in support of the nation's security. The proposals may focus on prototypes that offer potential for advancement and improvement of homeland security missions and operations.	n/a
Metropolitan Medical Response System Grant Program (FEMA)	This program provides funding to states to support the integration of local emergency management and medical systems into a coordinated local response system. Program funds can support purchasing of pharmaceuticals and personal protective equipment, among other things. This program has provided nearly $186 million to eligible states.	Fiscal years 2007-2011
Operation Stonegarden (FEMA)	This program provides funding to jurisdictions to enhance cooperation and coordination between law enforcement agencies that work to secure the U.S. borders. Funds must be used to improve coordinated operational capabilities of law enforcement agencies. This program has provided more than $234 million.	Fiscal years 2008-2011
Port Security Grant Program (FEMA)	This program's purpose is to protect critical port infrastructure from terrorism, particularly attacks that could cause a major disruption to commerce.	n/a
Port Security Grant Program–American	The American Recovery and Reinvestment Act of 2009 provided additional one-time	Fiscal years 2009

Table 6. (Continued)

Program	Description	Years funded (if available)
Recovery and Reinvestment Act funding (FEMA)	funding for this program to protect critical port infrastructure from terrorism.[b]	
State Homeland Security Program (FEMA)	This program awards grants to all 50 states, the District of Columbia, and 5 U.S. territories on the basis of risk and need. It provides funds to build state and local emergency response capabilities and implement state homeland security plans. This program has provided almost $3.1 billion to eligible entities.	Fiscal years 2008-2011
Tribal Homeland Security Grant Program (FEMA)	This grant program provides funds to eligible tribes for strengthening the nation against terrorism.	n/a
Trucking Security Program (FEMA)	This program supports the trucking industries' adoption and implementation of security measures, including global positioning systems tracking, and driver emergency alert notification systems, among other measures.	n/a
Urban Area Security Initiative Nonprofit Security Grant Program (FEMA)	This program provides funding to support nonprofit organizations located within a Urban Area Security Initiative region that are at high risk of a terrorist attack. Allowable costs include security communications and hardening activities.	n/a
Urban Areas Security Initiative (FEMA)	This program focuses on enhancing regional preparedness in major metropolitan areas.[c] It is intended to assist participating jurisdictions in developing integrated regional systems for prevention, protection, response, and recovery. This program has provided more than $3.8 billion.	Fiscal years 2007-2011

Source: DHS.

Note: Not available is referenced as n/a in the table.

[a] This program was jointly implemented by FEMA and OEC. FEMA defunded this program in fiscal year 2011 and instead incorporated the program's emergency communications goals and activities into the Homeland Security Grant program. Each of the five programs that comprise the Homeland Security Grant Program allows the grantee to purchase interoperable communications equipment.

[b] Pub. L. No. 111-5.

[c] FEMA has identified 31 highest risk urban areas eligible for Urban Areas Security Initiative funding.

Department of Justice

Two offices within Department of Justice, the Community Oriented Policing Services (COPS) and the Office of Justice Programs (OJP), administer grants that allow funds to be directed towards public safety emergency communications (see table 7).

Table 7. Department of Justice Grants with Emergency Communications as an Allowable Expenditure

Program	Description	Years funded (if available)
Edward Byrne Memorial Justice Assistance Grant (OJP)	This grant program supports many components of the criminal justice system, from multijurisdictional drug and gang task forces to crime prevention. In fiscal year 2011, the program awarded $360 million in grants for 1,400 grantees.	Fiscal years 2006-2011
Interoperable Communications Technology Grant Program (COPS)	Program provides funding for continued development of technologies and automated systems to help state, local, and tribal law enforcement agencies increase interoperability. The program awarded approximately $250 million to eligible law enforcement agencies.	Fiscal years 2003-2006
Law Enforcement Technology Program (COPS)	A noncompetitive program focused on the development of technologies to enable better response, investigation, and prevention of crime. The program provided $1.3 billion dollars grants to 18,000 local law enforcement agencies.	Fiscal years 1998-2010
National Institute of Justice Research Grants (OJP)	This program funded research in three areas: (1) enhancing the safety of criminal justice officers, (2) advancing use of geospatial technologies in law enforcement, and (3) modeling and simulation of technologies for virtual criminal justice training.	n/a
Tribal Resources Grant Program (COPS)	Program provides funding directly to federally-recognized tribal jurisdictions with established law enforcement agencies. It consists of two types of grants: (1) hiring grants and (2) equipment and training grants.	n/a

Source: DHS and the Department of Justice.

Note: Not available is referenced as n/a in the table.

Other Agency Programs

Six additional federal agencies administer grants that can fund public safety emergency communications, including the Departments of Agriculture (USDA), Transportation (DOT), Health and Human Services (HHS), Education, Interior, and the U.S. Navy (see table 8).

Table 8. Other Grants with Emergency Communications as an Allowable Expenditure

Program	Description	Years funded (if available)
Broadband Initiatives Program (USDA)	This program was funded by ARRA with the purpose of awarding grants and loans to facilitate broadband deployment in rural communities. This one-time program provided $3.5 billion in loans and grants for 320 projects.	Fiscal years 2009-2010
Communications and Networking (U.S. Navy)	This program provides funding for institutions and individuals' research and development of antennas, radio communications and wireless networking relevant to naval applications.	n/a
Enhanced 911 Grant Program (DOT)	This program provided grants to help 911 call centers implement next-generation technologies, such as the receipt of video or text messages from wireless callers or other features that could improve emergency response or enhance safety. The program provided more than $40 million in grants.	n/a
Hospital Preparedness Program (HHS)	The program supports the ability for hospitals and health care systems to prepare for and respond to bioterrorism and other public health emergencies. All awardees are required to equip participating healthcare entities with communication devices.	n/a
Public Health Emergency Response Grant Program (HHS)	The purpose of this grant was to support and enhance the state and local public health infrastructure that is critical to public health preparedness and response in the event of an influenza pandemic. Sixty-two entities, including the 50 states, 4 municipalities, and 8 territories and freely associated states were awarded this one-time grant.	n/a
Readiness and Emergency Management for Schools (Education)	Program provides funds to local educational agencies to establish an emergency management process that focuses on reviewing and strengthening emergency management plans.[a]	n/a

Program	Description	Years funded (if available)
Rural Development Community Connect Grant Program (USDA)	The program provides financial assistance to provide broadband service in rural communities without broadband. The grants establish broadband service for critical facilities, such as fire or police stations, while also providing service to residents and businesses.	n/a
Rural Development Community Facilities Programs (USDA)	This program provides grants and loans to rural public safety agencies by financing needed equipment, improvements, and services.	n/a
Rural Fire Assistance Outreach (Interior)	This program supports increasing local firefighter safety and enhancing fire protection capabilities of rural fire departments by providing basic wildland firefighting supplies and equipment.	n/a
State Health Information Exchange Cooperative Agreement Program (HHS)	This program provided funding to facilitate and expand the secure, electronic movement and use of health information among organizations according to nationally recognized standards that promote interoperability.	n/a

Source: DHS and other funding agencies.

Note: Not available is referenced as n/a in the table.

[a] The term "local educational agency" means a public board of education or other public authority legally constituted within a State for either administrative control or direction of, or to perform a service function for, public elementary schools or secondary schools in a city, county, township, school district, or other political subdivision of a state, or of or for a combination of school districts or counties that is recognized in a state as an administrative agency for its public elementary schools or secondary schools.

End Notes

[1] The radio frequency spectrum is the part of the natural spectrum of electromagnetic radiation lying between the frequency limits of 3 kilohertz (kHz) and 300 gigahertz (GHz). Radio signals travel through space in the form of waves. These waves vary in length, and each wavelength is associated with a particular radio frequency.

[2] Capacity refers to a communication system's ability to handle demand, provide coverage, and send different types of information.

[3] 9/11 Commission, *The 9/11 Commission Report: Final Report of the National Commission on Terrorist Attacks Upon the United States* (Washington, D.C.: July 2004).

[4] Congress, in the Balanced Budget Act of 1997, Pub. L. No. 105-33, 111 Stat. 251, mandated that FCC allocate 24 MHz of spectrum for public safety services by January 1, 1998.

[5] See Auction 73, 700 MHz Band, at (last accessed Feb. 17, 2012) http://wireless.fcc.gov/auctions/default.htm?job=auction_summary&id=73.

[6] As of February 17, 2012, the House of Representatives and the Senate had adopted the conference report accompanying the Middle Class Tax Relief and Job Creation Act of 2012, but the legislation had not been signed by the President at the time our work was completed on February 21, 2012. H.R. Rep. 112-399, accompanying the Middle Class Tax Relief and Job Creation Act of 2012, H.R. 3630, 112th Cong. (2012) as reported out on February 16, 2012.

[7] FCC, *A Broadband Network Cost Model: A Basis for Public Funding Essential to Bringing Nationwide Interoperable Communications to America's First Responders*, OBI Technical Paper No. 2 (May 2010). FCC staff told us they believe, based on information in the marketplace, that the cost of the network has risen since the May 2010 publication. The Middle Class Tax Relief and Job Creation Act authorizes $7 billion for the construction of the network from the potential proceeds of "incentive auctions." Incentive auctions are a special type of auction in which an existing user could receive a portion of the proceeds from the auction if the user relinquishes its rights to the spectrum.

[8] Public safety officials include all emergency responders and public safety agencies, such as firefighters, police officers, and paramedics who are the first to arrive at the scene of an emergency, as well as other responders such as hospital personnel, who might not be on the scene of an emergency but are essential in supporting effective response and recovery operations. Public safety agencies include 911 call centers that are also essential in supporting an effective response.

[9] NPSTC is a federation of organizations whose mission is to improve public safety communications and interoperability through collaborative leadership.

[10] The P25 suite of standards contains 8 open interfaces that exist between the various components of an LMR system. According to a PSCR official, of the 8 interfaces, only about 1.5 were complete at the time of our report.

[11] Homeland Security Act of 2002, Pub. L. No. 107-296, §502(5) and (6),116 Stat. 2135, 2212 (2002).

[12] The number of licenses excludes the 700 MHz public safety broadband license, the 4.9 GHz band, and public safety point-to-point microwave licenses.

[13] Pending legislation, the Middle Class Tax Relief and Job Creation Act of 2012, increases the amount of spectrum FCC is required to allocate for public safety from 24 to 34 MHz.

[14] The remaining 2 MHz were used as guardbands to protect from unwanted interference. *Second Report and Order*, 22 FCC Rcd. 15289 (2007).

[15] Spectrum is divided into frequency bands, each having technical characteristics that affect electronic transmission in different ways. "Bandwidth" is related to the transmission capacity of a frequency band and both the bandwidth and the frequency band can be described in MHz.

[16] FCC issued two further notices of proposed rulemakings since it attempted to auction the D Block, and a final order has not been adopted.

[17] When FCC auctions spectrum, the proceeds are to be deposited in the U.S. Treasury. 49 U.S.C. §309(j)(8).

[18] *Third Report and Order and Fourth Further Notice*, Service Rules for the 698-746, 747-762 and 777-792 Bands; Implementing a Nationwide, Broadband, Interoperable Public Safety Network in the 700 MHz Band, 26 FCC Rcd 733 (2011).

[19] *Requests for Waiver of Various Petitioners to Allow the Establishment of 700 MHz Interoperable Public Safety Wireless Broadband Networks*, Order, 25 FCC Rcd 5145, May 12, 2010; *Requests for Waiver of Various Petitioners to Allow the Establishment of 700 MHz Interoperable Public Safety Wireless Broadband Networks*, Order, 25 FCC Rcd 6783, May 12, 2011.

[20] LTE—the standard created and adopted by the Third Generation Partnership Project, a standards organization—is the closest standard to fourth generation wireless (4G)

technology that existed at that time. LTE has been accepted and adopted by national and international communities as the foundation for future mobile telecommunications.

[21] American Recovery and Reinvestment Act of 2009, Pub. L. No. 111-5, 123 Stat. 115 (2009).

[22] FCC issued its *Third Report and Order and Fourth Further Notice of Proposed Rulemaking* in this proceeding. See, Service Rules for the 698-746, 747-762 and 777-792 Bands; Implementing a Nationwide, Broadband, Interoperable Public Safety Network in the 700 MHz Band; Amendment of Part 90 of the Commission Rules, *Third Report and Order and Fourth Notice of Proposed Rulemaking,* 23 FCC Rcd 14301 (2011).

[23] Pub. L. No.109-295, §671, 120 Stat.1433, 1440 (2006).

[24] Funding from these grants can support emergency communications; not all funding was spent on the LMR voice communication systems.

[25] The Public Safety Interoperable Communications Grant Program is an NTIA program administered by the Federal Emergency Management Agency.

[26] Mississippi retained a portion of the grant funds for administrative purposes.

[27] FCC requires that public safety LMR systems migrate to at least 12.5 kHz efficiency technology by January 1, 2013, in an effort to ensure more efficient use of the spectrum. 47 C.F.R. §90.209(b)(5), fn. 3.

[28] In comparison, FCC has licensed about 260 MHz of spectrum for broadband personal communication services through auctions, which are a market-based mechanism in which FCC assigns a license to the entity that submits the highest bid for specific bands of spectrum. Public safety has received spectrum outside of the auction process.

[29] Approximately half of this allocation is located at 4.9 GHz, which according to DHS has limited value to public safety. Ideal spectrum for public safety lies in the 150 to 800 MHz bands.

[30] GAO, *First Responders: Much Work Remains to Improve Communications Interoperability*, GAO-07-301 (Washington, D.C.: Apr. 2, 2007).

[31] Establishment of an Emergency Response Interoperability Center, PS Docket 06-229, *Order*, FCC 10-67 (rel. Apr. 23, 2010).

[32] The Public Safety Advisory Committee is subject to the Federal Advisory Committee Act; the Technical Advisory Committee is not. The Federal Advisory Committee Act regulates the creation, operation, and termination of executive branch advisory committees.

[33] Federal users are allowed to access the public safety broadband network, subject to the public safety broadband licensee's approval, 47 C.F.R. § 2.103(c)(1). FCC cannot license federal users, but federal users might be able to subscribe to the network.

[34] The LTE standard is still being developed.

[35] LMR devices typically operate for more than 10 years, much longer than the standard commercial device.

[36] The technical design of a cellular network involves multiple hexagonal shaped cells merging together to form the network.

[37] One waiver jurisdiction ran simulations on 10 MHz of 700 MHz spectrum and determined that when overloaded the network became unusable for video and data.

[38] Pending legislation, the Middle Class Tax Relief and Job Creation Act of 2012, requires the creation of a Technical Advisory Board for First Responder Interoperability to develop recommended minimum technical requirements to ensure a nationwide level of interoperability.

[39] Pending legislation, the Middle Class Tax Relief and Job Creation Act of 2012, authorizes proceeds ($7 billion) from incentive auctions to construct the public safety network and authorizes the First Responder Network Authority to assess and collect fees to enable the authority to recoup its total expenses annually.

[40] The competitiveness of a market may be determined by factors other than the number of manufacturers within it—a market with only one manufacturer may be competitive if the manufacturer faces a credible threat of entry by competitors because the possibility that competitors will take away customers is enough to keep prices down.

[41] Encryption features convert data into a code to prevent unauthorized access. Although the P25 standard permits certain encryption features in handheld LMR devices, some device manufacturers sell encryption technology that falls outside the P25 standard.

[42] GAO.

[43] Further, incumbency advantages may be apparent in the standards development process. Stakeholders told us that because device manufacturers participate in the standards development process, the technology standards for handheld LMR devices may reflect the interests of dominant device manufacturers.

[44] GAO, *Transit Rail: Potential Rail Car Cost-Saving Strategies Exist*, GAO-10-730 (Washington, D.C.: June 30, 2010), and GAO, *DOD and VA Pharmacy: Progress and Remaining Challenges in Jointly Buying and Mailing Out Drugs*, GAO-01-588 (Washington, D.C.: May 25, 2001).

[45] GAO, *Information Assurance: National Partnership Offers Benefits, but Faces Considerable Challenges*, GAO-06-392 (Washington, D.C.: Mar. 24, 2006).

[46] According to DHS, equipment that is found compliant is posted to the Responder Knowledge Base at (last accessed, Feb. 17, 2012) https://www.rkb.us/.

[47] An alternative approach to fostering joint procurement is through a federal supply schedule. In 2008, the Local Preparedness Acquisition Act, Pub. L. No. 110-248, 122 Stat. 2316 (2008), gave state and local governments the opportunity to buy emergency response equipment through GSA's Cooperative Purchasing Program. The Cooperative Purchasing Program may provide a model for extending joint procurement to state and local public safety agencies.

[48] In two of the jurisdictions that received NTIA funding, we also met with technology vendors.

[49] It would be very difficult to review all grants to identify the total funding spent exclusively on emergency communications.

[50] Interoperable Communications Equipment eligible for purchase under DHS's authorized equipment list can be found at (last accessed Feb. 17, 2012) https://www.rkb.us/mel.cfm?expand=1&subtypeid=549. See Category 6 for Interoperable Communication Equipment.

INDEX

D

E

F

G

H

I

J

L

M

N

O

P

T

U

V

W